Advances in Anatomy, Embryology and Cell Biology
Ergebnisse der Anatomie und Entwicklungsgeschichte
Revues d'anatomie et de morphologie expérimentale

Springer-Verlag Berlin · Heidelberg · New York

This journal publishes reviews and critical articles covering the entire field of normal anatomy (cytology, histology, cyto- and histochemistry, electron microscopy, macroscopy, experimental morphology and embryology and comparative anatomy). Papers dealing with anthropology and clinical morphology will also be accepted with the aim of encouraging co-operation between anatomy and related disciplines.

Papers, which may be in English, French or German, are normally commissioned, but original papers and communications may be submitted and will be considered so long as they deal with a subject comprehensively and meet the requirements of the "Advances".

For speed of publication and breadth of distribution, this journal appears in single issues which can be purchased separately; 6 issues constitute one volume.

It is a fundamental condition that submitted manuscripts have not been, and will not simultaneously be submitted or published elsewhere. With the acceptance of a manuscript for publication, the publisher acquire full and exclusive copyright for all languages and countries.

25 copies of each paper are supplied free of charge.

Die Ergebnisse dienen der Veröffentlichung zusammenfassender und kritischer Artikel aus dem Gesamtgebiet der normalen Anatomie (Cytologie, Histologie, Cyto- und Histochemie, Elektronenmikroskopie, Makroskopie, experimentelle Morphologie und Embryologie und vergleichende Anatomie). Aufgenommen werden ferner Arbeiten anthropologischen und morphologischklinischen Inhalts, mit dem Ziel, die Zusammenarbeit zwischen Anatomie und Nachbardisziplinen zu fördern.

Zur Veröffentlichung gelangen in erster Linie angeforderte Manuskripte, jedoch werden auch eingesandte Arbeiten und Originalmitteilungen berücksichtigt, sofern sie ein Gebiet umfassend abhandeln und den Anforderungen der „Ergebnisse" genügen. Die Veröffentlichungen erfolgen in englischer, deutscher und französischer Sprache.

Die Arbeiten erscheinen im Interesse einer raschen Veröffentlichung und einer weiten Verbreitung als einzeln berechnete Hefte; je 6 Hefte bilden einen Band.

Grundsätzlich dürfen nur Arbeiten eingesandt werden, die nicht gleichzeitig an anderer Stelle zur Veröffentlichung eingereicht oder bereits veröffentlicht worden sind. Der Autor verpflichtet sich, seinen Beitrag auch nachträglich nicht an anderer Stelle zu publizieren.

Die Mitarbeiter erhalten von ihren Arbeiten zusammen 25 Freiexemplare.

Les résultats publient des sommaires et des articles critiques concernant l'ensemble du domaine de l'anatomie normale (cytologie, histologie, cyto- et histochimie, microscopie électronique, macroscopie, morphologie expérimentale, embryologie et anatomie comparée). Seront publiés en outre les articles traitant de l'anthropologie et de la morphologie clinique, en vue d'encourager la collaboration entre l'anatomie et les disciplines voisines.

Seront publiés en priorité les articles expressément demandés, nous tiendrons toutefois compte des articles qui nous seront envoyés dans la mesure où ils traitent d'un sejet dans son ensemble et correspondent aux standards des «Revues». Les publications seront faites en langues anglaise, allemande et française.

Dans l'intérêt d'une publication rapide et d'une large diffusion les travaux publiés paraitront dans des cahiers individuels, diffusés séparément: 6 cahiers forment un volume.

En principe, seuls les manuscrits qui n'ont encore été publiés ni dans le pays d'origine ni à l'éntranger peuvent nous être soumis. L'auteur s'engage en outre à ne pas les publier ailleurs ultérieurement.

Les auteurs recevront 25 exemplaires gratuits de leur publication.

Manuscripts should be addressed to/Manuskripte sind zu senden an/Envoyer les manuscrits à:

Prof. Dr. A. **BRODAL**, Universitetet i Oslo, Anatomisk Institutt, Karl Johans Gate 47 (Domus Media), Oslo 1/Norwegen

Prof. W. **HILD**, Department of Anatomy, Medical Branch, The University of Texas, Galveston, Texas 77550/USA

Prof. Dr. J. van **LIMBORGH**, Universiteit van Amsterdam, Anatomisch-Embryologisch Laboratorium, Mauritskade 61, Amsterdam-O/Holland

Prof. Dr. R. **ORTMANN**, Anatomisches Institut der Universität, Lindenburg, D-5000 Köln-Lindenthal

Prof. Dr. T. H. **SCHIEBLER**, Anatomisches Institut der Universität, Koellikerstraße 6, D-8700 Würzburg

Prof. Dr. G. **TÖNDURY**, Direktion der Anatomie, Gloriastraße 19, CH-8006 Zürich/Schweiz

Prof. Dr. E. **WOLFF**, Collège de France, Laboratoire d'Embryologie Expérimentale, 49 Avenue de la belle Gabrielle, Nogent-sur-Marne 94/Frankreich

Advances in Anatomy, Embryology and Cell Biology
Ergebnisse der Anatomie und Entwicklungsgeschichte
Revues d'anatomie et de morphologie expérimentale

53/1

Rolf Baur

Morphometry of the Placental Exchange Area

With 37 Figures

Springer-Verlag Berlin Heidelberg New York 1977

Dr. Rolf Baur, Anatomisches Institut der Universität Basel, Pestalozzistraße 20,
CH-4056 Basel, Switzerland

ISBN-13: 978-3-540-08159-3   e-ISBN-13: 978-3-642-66603-2
DOI: 10.1007/ 978-3-642-66603-2

Library of Congress Cataloging in Publication Data. Baur, Rolf, 1940-   Morphometry of the placental exchange area. (Advances in anatomy, embryology, and cell biology; 53/1) Bibliography:
p.   Includes index. 1. Placenta. 2. Morphology (Animals) 3. Maternal-fetal exchange. I. Title.
II. Series. QL801.E67 vol. 53/1 [QL973] 574.4'08s [599'.03'3] 77-3148

Composition: H. Stürtz AG, Universitätsdruckerei, Würzburg
2121/3321-543210

Contents

Symbols and Abbreviations

Stereology

S	Surface area
S_T	Total surface area (of the placental villi)
S_R	Relative surface area (villous surface area in cm² available for 1 cm³ of tissue)
S_M	Macroscopic surface area (of the chorionic sac)
V	Volume
V_T	Total volume (tissue volume of foetus and placenta)
V_C	Compact volume (placentomes, placental disc or placental girdle)
A	Development age
$S^{1/2}$	Square root of the surface area
$V^{1/3}$	Cube root of the volume
S_V	Surface density
V_V	Volume density

Statistics

N	Number of data pairs. With full term placentas the mean value of several placentas of the same animal species is considered as 1 value. With all stages of the cat, pig and rat the mean value of several placentas and foetuses of the same uterus is considered as 1 value
b	Regression coefficient b_{yx}
Conf. b	Confidence range of b, with the probability of error shown in brackets
a	Axis intercept a_{yx}. In linear regression a is the value of y at x = 0; in logarithmic regression a is the value of y at x = 1.
r	Correlation coefficient (product/moment coefficient)
Int. x	Intercept of the regression straight line on the x-axis.
tg	Tangent of the angle of slope of the regression straight line (orthogonal regression according to Sokal and Rohlf, 1969)

Symbols and Abbreviations

A. Introduction

Considering the placenta from the functional point of view as an exchange organ between mother and foetus, it is noteworthy that placentas of all types have a common structural principle in that all placentas have structures enlarging the surface area available for exchange, as is also the case with other organs with similar functions, such as the lungs, kidneys and intestine. It may therefore be assumed that a quantitative relationship must exist between the structure of the placenta and its function. Accurate information on this relationship could contribute to better understanding of the placenta and of its function. The study of this relationship requires quantitative physiological and morphological data.

Detailed data on quantitative morphology (morphometry) are available mainly for the human placenta. Even these data, however, are rather sparse as far as the early stages of development are concerned, because most studies relate to full term placentas (see Aherne, 1975; Bender et al., 1974; Laga et al., 1973; Geissler et al., 1972; Baur, 1972; Cattoor, 1967; Aherne and Dunnill, 1966). We know of only a few morphometric studies concerned with placentas of other animal species. These studies relate to investigations on the development of the placenta of sheep (Stegeman, 1974) and of cattle (Baur, 1972), and measurements carried out on full term placentas of the guineapig (Müller et al., 1967) and on full term placentas of various other species (Baur, 1973).

Bearing in mind that experimental studies on the physiology of the placenta have been carried out chiefly on placentas of experimental animals, information on the morphometric data of the various placenta types acquires increasing importance.

The following questions in particular require further study: 1. How large is the exchange surface area of the placenta at various stages of gestation? 2. How can the growth of this surface area be described in function of other data? 3. What is the quantitative relationship between the available surface area and the volume of tissue whose metabolic exchange this surface area must ensure? 4. What are the similarities and differences between placentas of various types?

This study investigates the growth of the exchange surface area in the macroscopic and in the visual microscopic observation range. The possibility of a further increase in this surface area by structures in the submicroscopic (electron-microscopic) range is not considered in detail in this study.

B. Materials and Methods

1. Materials

1.1. Development Stages

The comparative investigation of the embryonal development of the morphometric parameters, presented in this study, is based on one typical example each of six distinct structural types of placenta. These six types can be briefly characterised with the help of classifications dating back to Strahl (1906) and Grosser (1909) and of further morphological criteria, even though these classifications present certain problems on closer scrutiny (Portmann, 1938; Ludwig, 1968). These six types and their typical examples are as follows: 1) Discoid placenta, haemochorial: Man. 2) Discoid placenta, haemochorial labyrinth with several trophoblast layers: Rat. 3) Zonary placenta, endotheliochorial: Cat. 4) Cotyledonary placenta, epitheliochorial: Cow. 5) Diffuse placenta with well-developed dendritic villi, epitheliochorial: Horse. 6) Diffuse placenta with folds and stump-like villi, epitheliochorial: Pig.

The numbers of placental development stages investigated in each of the above six species are shown in Table 1 (datum N, Col. 1). With the multiparous species (cat, rat, pig), the data used for each development stage were mean values of several foetuses and placentas of the same uterus. With human placentas, the data on 17 stages in the first half of pregnancy were obtained by our own measurements, whereas the data on 21 further stages in the second half of pregnancy were calculated from data of Aherne and Dunnill (1966).

The age distributions of the development stages studied in the various species are shown in Figs. 1-17. The time spans covered from the earliest investigated stage to parturition, expressed in % of the full term gestation period, were as follows: Man 87 % (from the 5th week to birth); pig 82 %; cow 80 %; horse 71 %; cat 63 %; rat 32 %. The earliest development stages, during which the main nutritional mode differs from that with a fully formed placenta and which therefore necessitate other investigation techniques, are not included in this study so far.

1.2. Full Term Placentas

The full term placentas used for inter-specific comparisons belonged to 30 different species. The list of species is given in the legend of Fig. 37.

2. Contributors

The author wishes to thank the following persons and institutions who made this study possible by helping with the procurement of the investigated material:

2.1. Development Stages

Foetuses and placentas of cows and pigs were obtained from the municipal slaughterhouse of the City of Aachen (Director: Dr. med. vet. Bongartz; veterinaries: Dr. Benning, Dr. Schomee). Foetuses and placentas of cats were obtained from Dr. med. vet. Uehlinger (Veterinary Hospital, Basle) and Dr. med. vet. Mertens (Aachen). Most of the human placentas studied were provided by Dr. med. Wespi (Aarau Cantonal Hospital). Foetuses and placentas of the horse (Hungarian horses from the collection of Prof. Schauder, Giessen) were kindly provided by Prof. Schummer (Institute for Veterinary Anatomy of Giessen University).

10

2.2. Full Term Placentas

Full term placentas of various species (including elephant, rhinoceros, giraffe, etc.) were kindly made available by Prof. Dr. Lang (Director of the Basle Zoological Garden). I should like to thank his scientific assistants Dr. Wackernagel and Drs. med. vet. Folsch, Hoffmann and Ruedi, and the keepers, for help in collecting the material. Further full term placentas of various species were made available by Dr. Naaktgeboren and Dr. van Utrecht (Zoological Institute of the University of Amsterdam). A nearly full term placenta of the dolphin, with foetus, was made available by Dr. van Bree (Zoological Museum of the University of Amsterdam). A placenta of the aardvark (Orycteropus afer) was made available by Dr. Taverne of the Zoological Institute of the University of Amsterdam (see Taverne and Bakker-Slotboom, 1970). Two full term placentas of the llama were provided by Mr. Mussken (Zoological Garden of Aachen). A full term placenta of the horse (Westphalian thoroughbred) was made available by Mr. Lejeune (Aachen), further full term placentas of horses (and ponies) were provided by Dr. med. vet. Packbier (Aachen). Full term placentas of the dog (German sheep-dog) were provided by Dr. Ernst (Tropical Institute, Basle).

3. Nomenclature

To describe the measurement methods and their results it is necessary to define a few terms some of which are already in use in stereology whereas others had to be coined specially for this study. The notations used (shown in brackets) are based on English terminology, following the recommendation of the International society for Stereology (see Underwood, 1970; Weibel, 1969). These notations are defined on page 7 and are used in Figs. 1–37.

Total Volume. The total volume (V_T) is the volume of tissue the metabolic exchange of which must be ensured by the placenta. It includes the volume of the foetus and that of the placenta itself, excluding the amniotic and the allantoic fluid. The total volume is expressed in cm³, but volumes of 1000 cm³ and over are sometimes also expressed in kg (1 kg = 1000 cm³) for simplicity.

Total Villous Surface Area. The total area surface (S_T) is the total surface area of the placental villi available as exchange surface between mother and foetus. It is expressed in cm².

Relative Villous Surface Area. The relative villous surface area (S_R) is the total villous surface area divided by the total volume, both as defined above ($S_R = S_T/V_T$). Some authors (e. g., Weibel, 1969) also call this parameter the "specific surface". It is expressed in cm²/cm³.

Macroscopic Surface Area of the Chorionic Sac. The macroscopic surface area (S_M) is the surface area of the chorionic sac measurable by macroscopic methods without taking into account the surface area enlargements in the microscopic or the sub-microscopic domains.

Diffuse Placentas. Placentas the surface-enlarging structures (villi) of which are distributed diffusely over the whole chorionic sac (examples: equidae, camelidae).

Compact Placentas. Placentas the surface-enlarging structures of which are grouped in certain areas of the chorionic sac in the form of compact, macroscopically distinct features (girdle, disc, placentomes; examples: bovidae, carnivores). In this study the concepts of compact and diffuse placentas are used purely descriptively and are not to be confused with the concepts of "bulky" and "extended" placentas coined by Portmann (1938).

Compact Parts of a Placenta. These are the macroscopically distinct parts of a compact placenta (discs, girdles or placentomes) containing the surface-enlarging structures, as opposed to the chorion laeve and to the inter-cotyledonary area.

Surface Density. The surface density (surface area per unit volume, S_V) is a parameter expressing the surface-enlarging effect of the placental villi in compact placentas. It states how many cm² of villous surface area are contained on average in 1 cm³ of tissue of the compact parts of the placenta.

Volume Density. The volume density (volume per volume, V_V) is a parameter expressing the proportion of the volume of the compact parts of the placenta composed of foetal tissue as opposed to maternal tissue.

Surface Enlargement Factor. The surface enlargement factor (f) expresses the surface-enlarging effect of the placental villi in diffuse placentas. This factor states the extent by which the macroscopically measurable surface area of the chorionic sac is increased by the presence of villi, i. e., it is the ratio, for diffuse placentas, of the total surface to the macroscopic surface, both as defined above ($f = S_T/S_M$).

4. Measurement Methods

The methods used for stereological measurements on placentas, and the associated problems, have been described in detail in a previous publication (Baur, 1973). The following text therefore presents only a brief survey of the requisite data and of the general procedure. The section on evaluation is treated in greater detail.

4.1. Macroscopic Data

Volumes of Foetuses. The volumes were determined by the water displacement method. Only when this was impossible (neonates) was the weight used instead of the volume. With multiparous species (several foetuses per uterus: cat, rat, pig) the mean values were used. With these species, each point in the diagrams represents the mean value of measurements on several foetuses and placentas of the same uterus.

Volume of the Placenta. With diffuse placentas the placenta was separated from the uterus and the volume of tissue, i. e., excluding the amniotic and allantoic fluid, was determined by the water displacement method.

Volume of the Compact Parts of the Placenta. With placentas of the compact type the volume of the compact parts of the placenta and that of the intercotyledonary area were determined separately. For this purpose, the compact parts of the placenta, e. g., the placentomes of the cow, were separated from the uterus and from other parts of the placenta. To calculate the volume of the foetal moiety of the placenta, the maternal tissue components contained in the compact parts must be subtracted (the volumetric proportion is determined stereologically).

Macroscopic Chorionic Surface Area. This area was determined on the placenta spread out flat, with the aid of a superimposed point grid, i. e., in accordance with the stereological principle of point counting method (Underwood, 1970; Weibel, 1969). The necessary corrections have been discussed by Baur (1973). Pig placentas were spread out in such a manner as to smooth out the macroscopically visible folds. With this procedure, the surface area enlargement due to these folds is included in the macroscopic chorionic surface area, so that the microscopically determined enlargement factor (f) encompasses in this case only the surface area enlargement due to the microscopic folds and the stump-like villi.

Ages of the Foetuses and Placentas. With rats, the ages of the embryos and foetuses were known because the rats were mated specially for these experiments. The ages of the embryos and foetuses of the other species studied were estimated on the basis of size, weight and external appearance according to published standard tables (Jakobovits et al., 1972; Marrable, 1971; Michel, 1968; Needham, 1963; Naakgeboren, 1960; Zietzschmann and Krölling, 1955; Postma, 1947; Stoss, 1944; Streeter, 1920; Gurlt, 1873).

4.2. Histological Preparation Procedure

The histological preparation procedure was as follows: 1. Taking *tissue samples* of each placenta. The selection of samples had to comply with the rules of representative sampling. 2. *Fixation:* Bouin, Carnoy. Material already fixed with formol was post-fixed with Bouin's solution. 3. *Embedding:* in paraffin, with methyl-benzoate and benzol as intermediates. Also used was embedding in polyethylene-glycol (Barka and Anderson, 1965). 4. *Section thickness:* 7 microns. 5. *Stains*: Azan, PAS, haemalum-Benzopurpurin. 6. *Frozen sections:* When fresh placentas were available, frozen sections were also prepared and stained with haemalum-Benzopurpurin and haemalum-erythrosin.

4.3. Microscopic Data

The requisite microscopic data were obtained from the histological sections by means of stereological methods (see Underwood, 1970; Weibel, 1969). Provided the histological sections used constitute statistically representative samples of the tissue to be studied, stereological methods make it possible to draw from two-dimensional section aspects information on quantitative relationships in three-dimensional space.

Volume Density (V_V). With placentas of the compact type (e. g., cow) it must be determined which proportions of the volume of the compact parts of the placenta, i. e., for example of the placentomes of the placenta of the cow, are formed by foetal and which by maternal tissue. In the section aspect the volume of the villi is represented by the cross-section areas of these villi. The measuring technique is based on superimposing on the histological section (or on a projection of this section) a measuring grid consisting of a known number of points arranged in a regular pattern at a known spacing. The points of the measuring grid coinciding with section areas through the villi are counted. This number is called the number of hits. According to the basic equations of stereology (Underwood, 1970; Weibel, 1969), the ratio of this number to the total number of points of the measuring grid is proportional to the volume density (V_V), i. e., to the percentile proportion of the volume of villi in the whole of the tissue studied.

Surface Density (S_V). The surface area of the villi is represented in the section aspect by the boundary lines of the cross-sections of the villi. The measuring technique is based on superimposing on the histological section (or on a projection of this section) a measuring grid consisting of lines of a known length and at a known uniform spacing, and counting the number of intersections between these grid lines and the boundary lines of the villi. From this number and from the known geometric data it is then possible, by means of the basic equations of stereology (Underwood, 1970; Weibel, 1969), to determine the surface density (S_V), i. e., the surface area of the villi (in cm²) contained in 1 cm³ of placental tissue. In placentas of the cow we determined, instead of the surface area of the villi, the surface area of the maternal crypts because it is less affected by the shrinkage of the tissue (Baur, 1973). With rat placentas we injected Indian ink into the blood vessels on the maternal side and we measured the surface area of the trophoblast adjoining the maternal blood spaces of the placental labyrinth.

Surface Enlargement Factor (f). For determining the surface enlargement factor of diffuse placentas a supplementary procedure was developed starting from the basic equations of stereology. The problem is to determine the factor (f) by which the surface area of a given basal area of the chorionic sac is enlarged by the presence of villi. The histological section are taken normally to the basal surface of the placenta. In these sections the basal area is represented by the length of the base line of the visual field, and the area of the villi is represented by the aggregate length of boundary lines of the villi seen in the section. The length of the base line is measurable directly, the aggregate length of the boundary lines of the villi can be determined as for (S_V) by counting the number of intersection points between these lines and those of the measuring grid. According to the basic equations of stereology, the quotient of the aggregate length of these boundary lines divided by the length of the base line corresponds to the surface enlargement factor, i. e., to the quotient of the villous surface over a given basal area divided by the size of this basal area (Baur, 1973). With pig placentas the surface enlargement factor was determined chiefly on frozen sections, because of the lesser extent of surface shrinkage.

4.4. Calculation of the Parameters

Total Volume. The total volume (V_T) is calculated as the sum of the volume of the foetus and the volume of the foetal components of the placenta (without amniotic and allantoic fluid). In compact placentas the volume of the foetal components of the placenta is calculated by subtracting the proportion of maternal tissue from the total volume of the compact parts of the placenta.

Total Villous Surface Area (S_T). With diffuse placentas, this area is obtained by multiplying the macroscopic chorionic surface area (S_M) by the surface enlargement factor (f). With compact placentas, this area is obtained by multiplying the volume of the compact parts of the placenta (V_C) by the surface density (S_V). With compact placentas, the areas located outside the compact parts of the placenta, i. e., the intercotyledonary area, the paraplacental area and the chorion laeve, are ignored in these calculations.

Relative Villous Surface Area (S_R). This area is obtained by dividing the total villous surface area (S_T) by the total volume (V_T).

5. Evaluation

5.1. Graphical Presentation

In order to study the interdependence of the parameters studied, the measured parameters were entered into two-parameter coordinate grids, with one of the two selected parameters plotted along the axis of ordinates and the other plotted along the axis of abscissae. Of the large numbers of possible two-parameter combinations, 14 combinations were selected which are relevant for the problems investigated in this study (Figs. 1–37). The following graphical presentations of these combinations were found useful:

Linear Coordinates (Both Axes). For each combination of two parameters, the two scales were so chosen that the range of the available data covered about the same length measured along either coordinate axis. This presents the advantage that deviations of the distribution of the data points from a straight line stand out particularly clearly. The drawback of this presentation method is that direct visual comparisons of the graphs for different species are made difficult by the unequal scales used. This drawback was considered acceptable because the comparisons were based on statistical methods.

Logarithmic Coordinates (Both Axes). One of the advantages of this presentation is that it makes it possible to show a range of several orders of magnitude on the same diagram (see Fig. 37).

Root Transformations. It is sometimes desirable to represent a surface area in terms of its square root, and a volume in terms of its cube root. This transformation converts an area into a square and a volume into a cube, and makes it possible to study the relationship of the corresponding linear dimensions, i. e., of the length of side of the area square and the length of edge of the volume cube.

5.2. Choice of Smoothing Curves

In those cases where the distribution of data points in a graphical presentation indicated a definable trend, it was endeavoured to fit this distribution with a smoothing curve amenable to mathematical definition by means of an equation. As all such curves presented in this study represent purely empirical functions and not a "growth law", we placed priority on simplicity, i. e., amongst the several mathematical equations which might have been used, we chose the simplest equations statisfying the following criteria: 1. The equation must represent the available data points with an accuracy sufficient for the purposes of our study; 2. The equation must provide meaningful biological information; 3. The equation must be amenable to statistical evaluation.

The simplest case arises when the data points can be represented by a straight line in either linear or logarithmic coordinates. In linear coordinates a straight line can be expressed by a linear equation of the general form

$$y = a + bx \tag{1a}$$

and in logarithmic coordinates it can be expressed by a power function of the general form

$$y = ax^b \tag{2a}$$

In most of the problems investigated in this study, one or the other of these two equations satisfies the criteria defined above and has therefore been used for the summary representation of various relationships. On the other hand, the use of these simple equations limits the deducible information to the numerical range covered by the measured data. If extrapolations beyond this range are considered necessary, then other equations must be used such as, for example, differential equations (Scharf, 1971), the logistic growth function (Kretschmann and Wingert, 1969) or the growth equation according to Bertalanffy (1957, 1951). A summary review of the various possibilities is given by Peil (1974). In our opinion, however, the use of such equations would be meaningful only if the course of the curves near the extreme ends of their range were substantiated more accurately by further measurements.

5.3. Calculation of the Smoothing Curves

The parameters of the smoothing curves for the linear equation (1a) were calculated by means of linear regression (regression of y on x, Sachs, 1972; Berkson, 1950), those for the power function (2a) were calculated by means of logarithmic regression with logarithmically transformed values. Bearing in mind that, in our study, both variables are burdened with errors, the general trends of the data points would have been defined somewhat more accurately with the aid of orthogonal regression (Sokal and Rohlf, 1969) or of allometric linear regression (Teissier, 1948) than by classical regression calculations. As a result, the values shown in Tables 1–7 are somewhat too low where they show a rising trend. However, as the resulting errors are small (see Table 5, Columns 2 and 8) and are not of great importance as far as the aims of this study are concerned, we gave preference to the classical regression calculations because they are more suitable for statistical evaluations.

The smoothing lines and their statistical parameters were computed on a programmable desk computer (Hewlett-Packard 9810 A). In these calculations, all the areas were expressed in cm², the

14

volumes in cm³ and the ages in weeks. In the values shown in Tables 1–7 the last digit after the decimal point is rounded off. In the case of the horse the values relating to full term placentas were not included in the same calculations as those relating to the development stages because the conditions of measurement were not the same in that, with the full term placentas, the macroscopic parameters could be measured in the fresh condition whereas the development stages were available only after long periods of fixation in formol.

5.4. Biological Interpretation of the Regression Parameters

We present below a brief discussion of the biological interpretation of the parameters used in Eqs. (1a) and (2a) when these equations are applied to studying the growth of the total villous surface area (S_T) and of the total volume (V_T) in function of age. It should be noted that the parameter designations a and b are used in this study in the sense of statistical regression calculations (see Sachs, 1972), i. e., in a different sense from that used in the allometric equation according to Teissier (1948).

x and y are the variables the interdependence of which is to be studied. In the case discussed here, y represents in linear coordinates the square root of the total villous surface area (S_T) or the cube root of the total volume (V_T) plotted as ordinates, and x represents the age (A) plotted as abscissae. In logarithmic coordinates y represents the total villous surface area and x the total volume.

Parameter a in Eq. (1a). In linear coordinates, the y-axis intercept a represents geometrically the intersection point of the regression straight line with the y-axis at the abscissa x = 0. A parameter of more direct importance from the biological point of view is the intercept of this straight line on the x-axis (point y = 0; x = −a/b), because this point represents the time of *beginning* of growth (assuming that the choice of a straight line as a model of growth is justified).

Parameter b in Eq. (1a). The regression coefficient b of linear regression corresponds to the tangent of the angle of slope of the regression straight line, i. e., it is a measure of the steepness of this slope. Biologically, in the case considered here, this coefficient is a measure of the *rate* of growth.

Parameter a in Eq. (2a). In logarithmic coordinates (both axes on a logarithmic scale), the y-axis intercept a represents the value of the variable y at the abscissa x = 1.

Parameter b in Eq. (2a). The regression coefficient b of logarithmic regression (exponent of the power function) represents the tangent of the angle of slope of the regression straight line plotted in logarithmic coordinates. Biologically, in the case considered here, this coefficient is a measure of the *acceleration* (or of the slowing down) of growth with time.

5.5. Assessment of the Significance of Differences

The significance of differences between statistical parameters was tested with the aid of confidence ranges because this allows a better assessment of the procedure than the usual statistical tests which are often used routinely. The width of the confidence range varies depending on the permitted probability of error. With the same set of data, a permitted probability of error of 0.1 (i. e., 10 %) results in a relatively narrow confidence range, whereas one of 0.01 (i. e., 1 %) results in a wider confidence range. In order to avoid statistical assessment errors of the first and of the second kind, i. e., the risk of declaring an actually existing difference as being statistically nonsignificant or of declaring a practically insignificant difference as being statistically significant, the following procedure was adopted:

A difference between two parameters was accepted as statistically significant only if the wide 1 % confidence ranges did not overlap at all. This finding should be formulated more precisely as follows: "Assuming that the investigated values are distributed normally, the difference may be considered significant with an error probability of 1 %". On the other hand, equality between two parameters was presumed as a rule only if even the narrow 10 % confidence ranges of the two parameters overlapped to some extent. This finding should be formulated more precisely as follows: "Assuming a normal distribution, no significant difference can be proven on the basis of the available values and extent of the sample".

A similar procedure was adopted for comparing regression straight lines as a whole. For these comparisons we used 0.05 confidence bands (i. e., error probability 5 %) of the regression straight lines, calculated by the formula according to Sachs (1972, 3rd ed., p. 342). These confidence bands, based on the 2F value, are wider than those based on the t value. A difference in level, i. e., in the direction of the y-axis, was considered to be significant only if the corresponding confidence bands did not overlap at all anywhere in the entire investigated range.

C. Findings

1. Growth of the Total Villous Surface Area

If we plot the measured values of the total villous surface area (S_T) against the development age, in linear coordinates, we obtain a family of data points rising towards the right (Figs. 1–4). In other words, in all the species studied, the total villous surface

Figs. 1–4. Growth of the total villous surface area (S_T). The smoothing curves are concave upwards, i. e., the rate of growth of the total villous surface area increases in the course of the development.

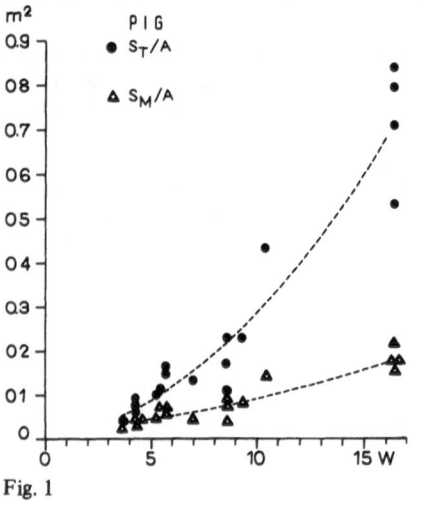

Fig. 1

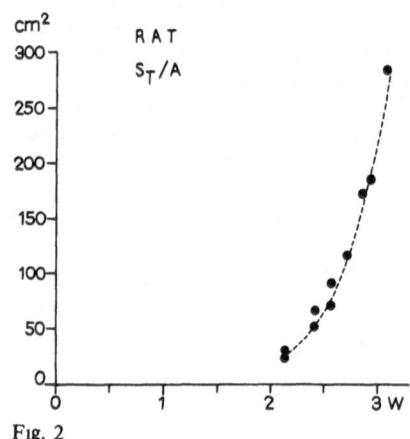

Fig. 2

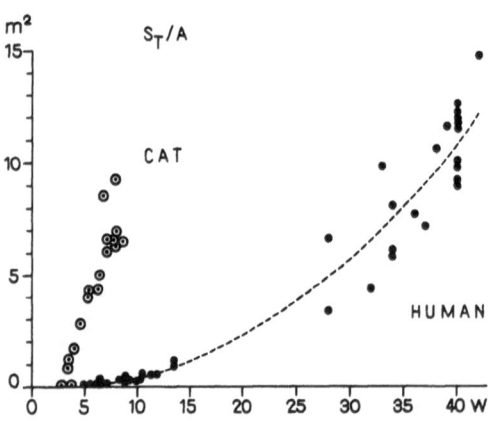

Fig. 1. Pig. In addition to the total villous surface area (S_T), Fig. 1 also shows the growth of the macroscopic surface area of the chorionic sac (triangles)

Fig. 2. Rat

Fig. 3. Cat (dotted circles) and man (full circles). In the case of the human placenta, the group of data points on the left originates from our own measurements, whereas the data points on the right (older stages) were calculated from the data of Aherne and Dunnill (1966)

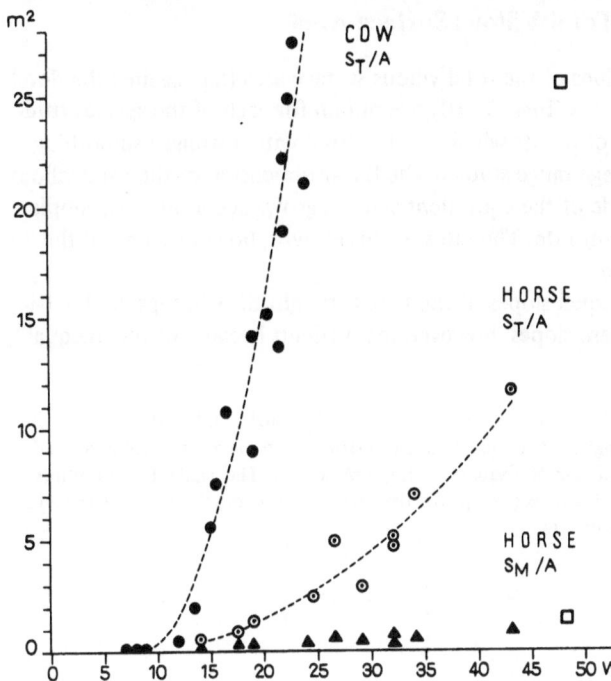

Fig. 4. Cow (full circles) and horse (dotted circles). The S_T curve for the cow placenta is steeper than that for the horse placenta (faster growth). For the cow, the S_T values for the older stages (36 to 120 m²) are off the scale in this diagram, but they continue the same trend (see Figs. 9 and 27). The data for the horse also show the growth of the macroscopic chorionic surface area (S_M) (triangles). The squares show mean values of 5 full term placentas of the horse (top square: S_T; bottom square: S_M)

area increases in the course of development of the placenta. The general trend of the data points can be represented in linear coordinates by a fitted parabolic smoothing curve. In other words, the rate of growth of the total villous surface area increases with time (acceleration). This applies to all the species studied, except the cat (Fig. 3). Figs. 3 and 4 each show the data for two species, plotted in the same coordinate system. The unequally steep rises of the curves show that the rate of growth of the total villous surface area is not the same in all species. For example, the overall growth is faster in the cow than in the horse (Fig. 4). These graphs also give a rough idea of the total villous surface area of full term placentas, e. g., about 12 m² in humans (Fig. 3). In the case of the cow the graph had to be cut short in order to fit it into the same diagram as the data for the horse (Fig. 4), but the size of the total villous surface area of the full term placenta can be read off Figs. 27 and 37 (about 130 m²). It can thus be seen that considerable inter-specific differences also exist in respect of the total villous surface area of the full term placenta.

2. Transformed Values of Areas and Volumes

The relationship between the growth of the total villous surface area (S_T) and that of the total volume (V_T) is difficult to study directly because the area grows two-dimensionally whereas the volume grows three-dimensionally. It is therefore handier to reduce the area and the volume to their respective linear dimensions by using as parameters the square root of the area and the cube root of the volume, respectively. Mathematically speaking, these parameters represent the length of side of a square with the same area as (S_T) and the length of edge of a cube with the same volume as (V_T), respectively. We can then study the growth of these two lengths. This results in easier visualisation of relationships because both these parameters are unidimensional.

2.1. Square Root of the Total Villous Surface Area

If we plot the square root values of the total villous surface area (S_T) against the development age in linear coordinates (Figs. 5–10), we obtain for each of the species studied (except the cat) a family of points which can be fitted with a straight smoothing line. This means that, in the age range studied, the linear dimension of the total villous surface area (the length of side of the equivalent square) grows according to a simple principle, namely, at a constant rate. The rates of this growth, however, are not the same in all the species studied.

This is reflected in the unequal slopes of the fitted straight lines in Figs. 5–10. Direct visual comparisons of these slopes, however, are difficult because of the unequal

Figs. 5–10. Square root values of the total villous surface area (S_T) (dots) and of the macroscopic chorionic surface area (S_M) (triangles). The square root transformation makes it possible to study the rate of growth of the length of side of a square with the same area. The available data points can be fitted with straight smoothing lines (except for the cat), i. e., the length of side of the equivalent square increases at a constant rate.

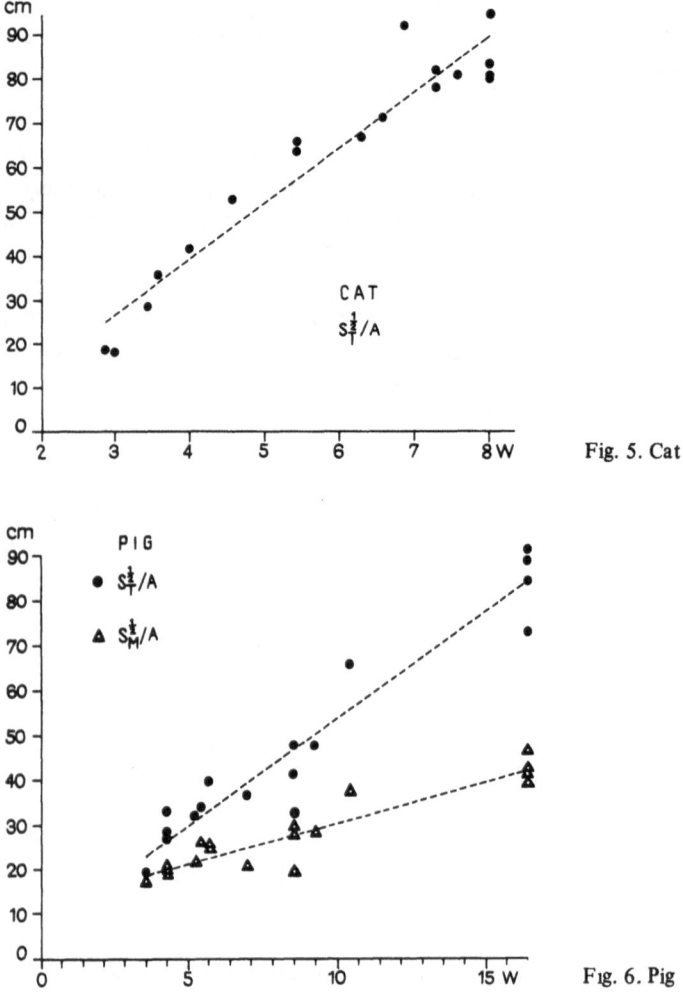

Fig. 5. Cat

Fig. 6. Pig

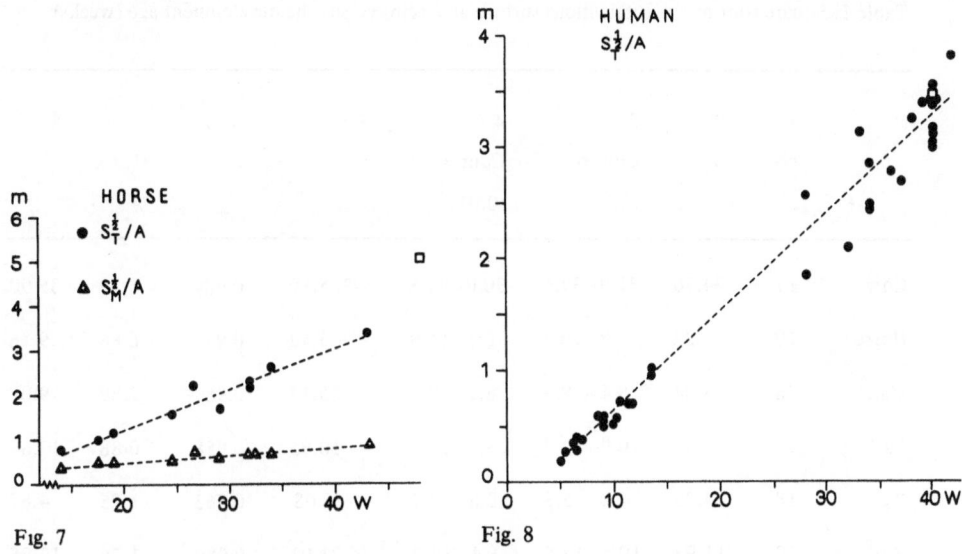

Fig. 7

Fig. 8

Fig. 7. Horse. The square shows the mean value of 5 fresh full term placentas. The other values are located relatively lower on the diagram because they have not been corrected for shrinkage caused by fixation

Fig. 8. Man. Left group of points: our own data. Right group (older stages): values calculated from the data of Aherne and Dunnill (1966). The square at 40 weeks represents the mean value of our own measurements on full term placentas

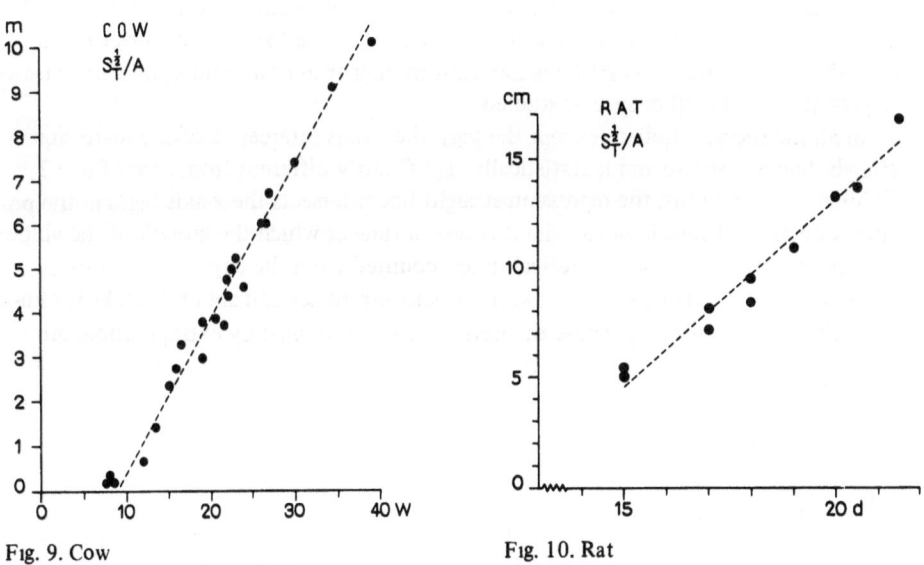

Fig. 9. Cow

Fig. 10. Rat

scales used in these diagrams. The differences in the rates of growth stand out more clearly if we compare the regression coefficients b (tangent of the angle of slope of the regression straight line, Table 1, Column 2). Columns 3 and 4 in Table 1 show the narrow (10 %) and the wide (1 %) confidence ranges of the linear regression coefficients

19

Table 1. Square root of the total villous surface area (cm) versus the development age (weeks)

	1	2	3	4	5	6	7	8
	N	b	Conf.b (0.1)	Conf.b (0.01)	a	r	Int.x	tg
Cow	17	34.76	31.9–37.6	30.0–39.5	−305.59	0.984	8,79	35.90
Horse	10	9.22	7.8–10.7	6.6–11.9	− 63.42	0.971	6.88	9.76
Man[a]	38	8.84	8.4– 9.3	8,2– 9.5	− 25.54	0.987	2.89	9.08
Cat[b]	18	12.51	10.9–14.2	9.8–15.3	− 10.95	0.958	0.88	13.64
Pig	18	4.76	4.2– 5.3	3.8– 5.7	6.08	0.963	−1.28	4.87
Rat[b]	10	11.93	10.4–13.5	9.1–14.7	− 21.01	0.980	1.76	12.28

[a] Including data of Aherne and Dunnill (1966)
[b] Linear regression as a simplification

b, by means of which it can be tested whether the observed differences in the rates of growth should be considered coincidental or statistically significant. It can be seen, for example, that the rate of growth of the square root of the total villous surface area (and therefore of that area itself) is significantly higher in cows, and significantly lower in pigs, than in the other species studied.

In all the species studied (except the pig), the y-axis intercept a of the regression straight line is negative and is statistically significantly different from zero (Table 1, Column 5). As a result, the regression straight line intersects the x-axis (age) in the positive domain and thus indicates the theoretical time at which the growth of the villous surface area begins. These theoretical times, counted from the day of conception, come to about 9 weeks for cows, about 7 weeks for horses and about 3 weeks for humans (Table 1, Column 7). These theoretical values, obtained by extrapolation, are

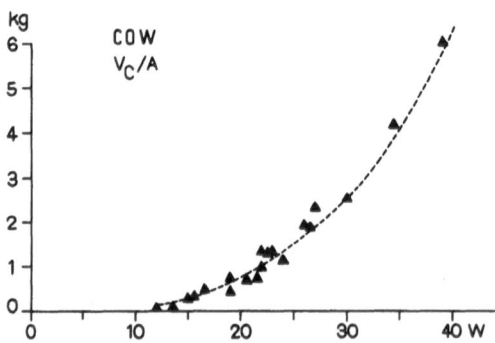

Fig. 11. Volumetric growth of the compact parts of the placenta (V_C, in this instance: placentomes of the cow). The smoothing curve (section of a cubic parabola) shows an increasing rate of volumetric growth in the course of development

more or less in agreement with the actual biological conditions. In the case of pigs, however, the regression straight line intersects the x-axis in the negative domain, which is biologically meaningless. This means that, in this particular case, extrapolations beyond the time range studied are not permissible.

2.2. Cube Root of the Total Volume

If we plot the cube root values of the total volume (V_T, volume of the foetus plus the foetal components of the placenta) against the development age in linear coordinates we obtain, for all the six species studied, families of data points rising towards the right, which shows that the total volume increases in the course of development. With all the six species studied, these families of data points can be fitted with straight smoothing lines (Figs. 12–17). This shows that the volumetric growth rate, expressed in terms of the length of edge of the equivalent cube, is constant for each species within the development age range covered by the measurements. These rates of growth, however, are not the same in all the species studied.

Figs. 12–17. Cube root values of the total volume (V_T) (circles) and of the volume of the compact parts of the placenta (V_C, for compact placentas only) (triangles). The cube root transformation makes it possible to study the rate of growth of the length of edge of a cube with the same volume. The available data points for V_T can be fitted with straight smoothing lines for all the species studied, showing that, within the development period studied, the length of side of the equivalent cube increases at a constant rate

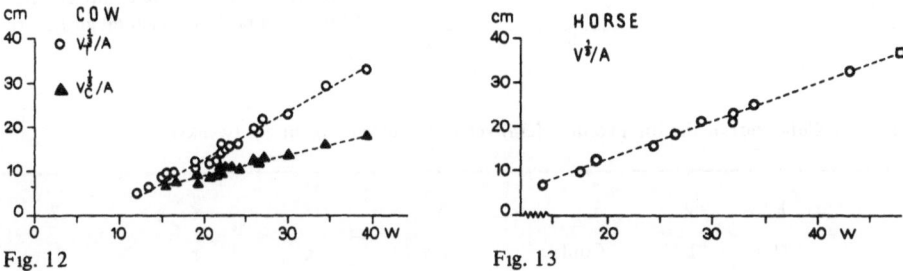

Fig. 12   Fig. 13

Fig. 12. Cow. The curved smoothing line for V_C in Fig. 11 is converted by the cube root transformation into a straight line (triangles)

Fig. 13. Horse. The square represents the mean value of 5 fresh full term placentas

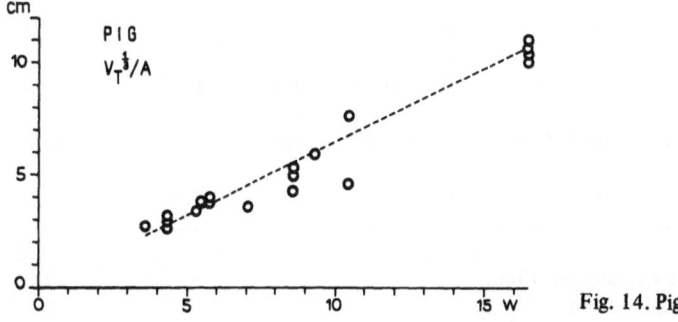

Fig. 14. Pig

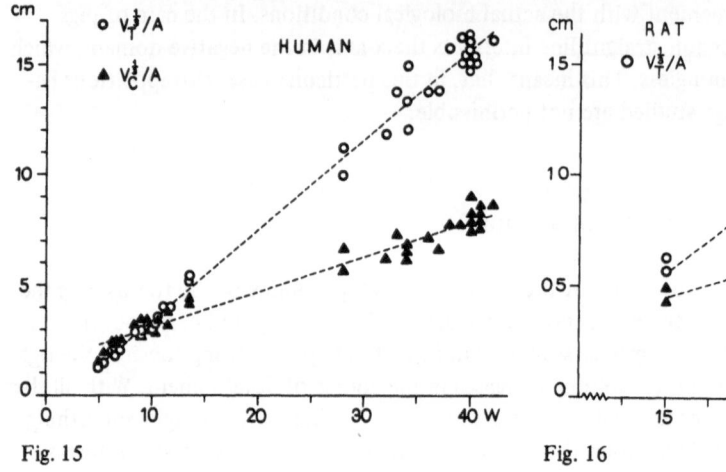

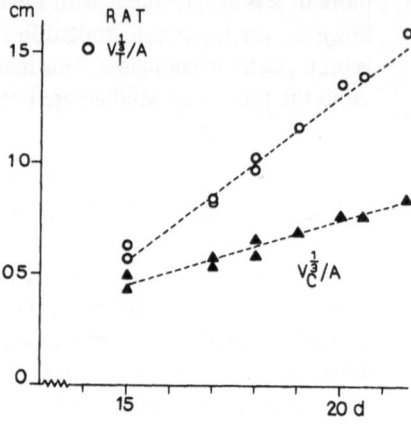

Fig. 15

Fig. 16

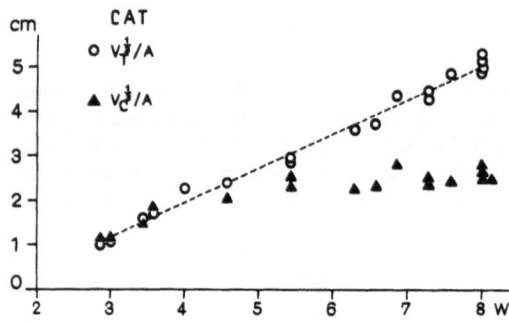

Fig. 15. Man. Left group of symbols: our own data. Right group (older stages): values calculated from the data of Aherne and Dunnill (1966)

Fig. 16. Rat

Fig. 17. Cat. A straight smoothing line for the transformed values is permissible for the total volume (V_T, circles), but not for the volume of the compact parts of the placenta (V_C, placental girdle, triangles)

Table 2. Cube root of the total volume (cm) versus the development age (weeks)

	1	2	3	4	5	6	7
	N	b	Conf.b (0.1)	Conf.b (0,01)	a	r	Int.x
Cow	17	1.08	1.01–1.16	0.95–1.22	−9.21	0.987	8.49
horse	10	0.87	0.81–0.93	0.76–0.98	−5.02	0.994	5.76
Man[a]	38	0.41	0.40–0.42	0.39–0.43	−0.71	0.996	1.73
Cat[b]	18	0.77	0.74–0.81	0.71–0.83	−1.12	0.994	1.45
Pig	18	0.65	0.61–0.70	0.58–0.73	−0.007	0.987	0.01
Rat[b]	10	1.06	0.99–1.13	0.93–1.19	−1.70	0.994	1.61

[a]Including data of Aherne and Dunnill (1966)
[b]Linear regression as a simplification

22

This can be seen most clearly by comparing the unequal linear regression coefficients b (Table 2, Column 2). It can be seen, for example, that of the six species studied, the rate of growth of the cube root of the total volume (and therefore of that volume itself) is lowest for humans and highest for cows. The statistical significance of the differences can be tested with the aid of the confidence ranges of the regression coefficients (Table 2, Columns 3 and 4). Thus, comparing man and cow, it must be concluded that the difference is statistically significant because there is no overlap even between the wide (1 %) confidence ranges (Column 4). Conversely, comparing cow and rat, no statistically significant difference can be proven on the basis of the available data, because even the narrow (10 %) confidence ranges (Column 3) overlap to a considerable extent.

The regression straight lines intersect the x-axis (age) in the positive domain. In the cases of cow, horse and man, however, the theoretical starting times of volumetric growth defined by these intersection points (Table 2, Column 7) are about one week earlier than those based on areal growth (Table 1, Column 7).

3. Variation of the Area/Volume Ratio with Time

3.1. Linear Coordinates

If we plot the measured values of the total villous surface area (as ordinates) against the corresponding measured values of the total volume (as abscissae) in linear coordinates, we obtain for each of the six species studied a family of data points which can be fitted with a smoothing curve convex upwards (Figs. 18–23). The curve begins at

Figs. 18–23. Total villous surface area (S_T) plotted against the total volume (V_T) during development. The smoothing curves are concave downwards, i. e., the area/volume ratio decreases in the course of development, so that the villous surface area available per unit of tissue volume is smaller for older foetuses than it is in the earlier stages

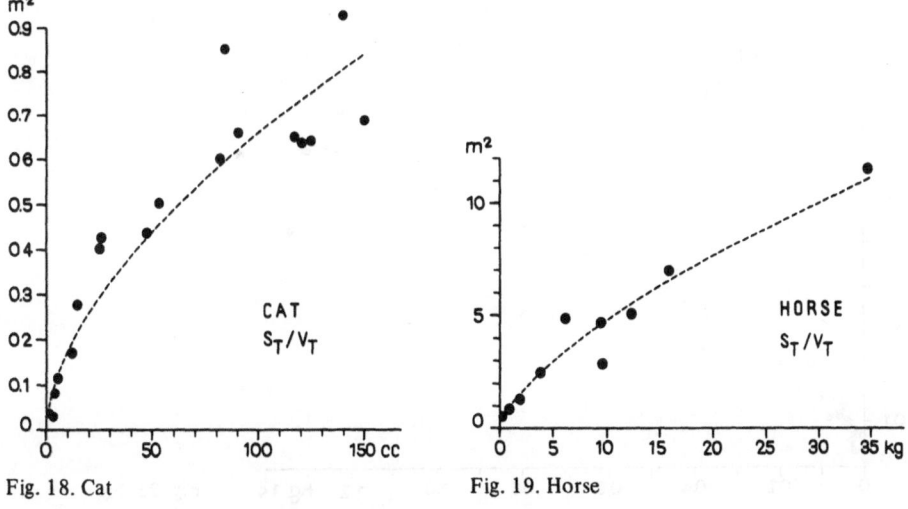

Fig. 18. Cat   Fig. 19. Horse

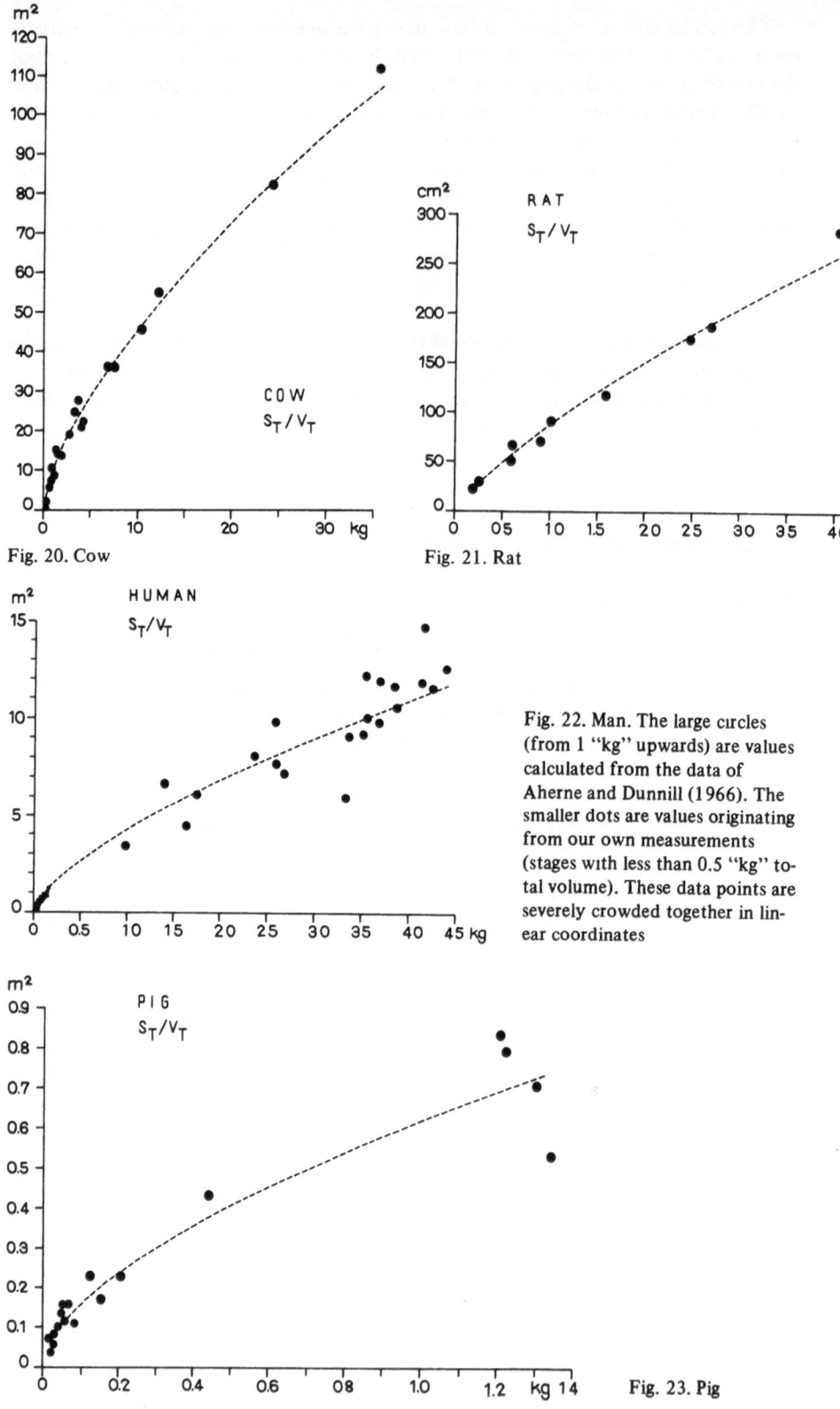

Fig. 20. Cow

Fig. 21. Rat

Fig. 22. Man. The large circles (from 1 "kg" upwards) are values calculated from the data of Aherne and Dunnill (1966). The smaller dots are values originating from our own measurements (stages with less than 0.5 "kg" total volume). These data points are severely crowded together in linear coordinates

Fig. 23. Pig

Table 3. Square root of the macroscopic chorionic surface area (horse, pig) and cube root of the volume of the compact parts of the placenta (cow, man, cat, rat), versus the development age (weeks)

	1 N	2 b	4 Conf.b (0.01)	5 a	6 r
Horse	10	1.65	0.90–2.39	15.1	0.934
Pig	18	1.84	1.31–2.37	12.1	0.930
Cow	17	0.49	0.42–0.56	−0.7	0.982
Man[a,b]	38	0.16	0.15–0.17	1.4	0.982
Cat[b]	17	0.25	0.16–0.34	0.7	0.899
Rat	10	0.41	0.30–0.52	−0.4	0.977

[a] Including data of Aherne and Dunnill (1966)
[b] Linear regression as a simplification

the origin of the coordinate systems and rises steeply at first and progressively less steeply later. The shape of the smoothing curve shows that, at first, the total villous surface area grows relatively faster than the total volume which it serves. As development progresses, however, the relative growth rate of the total volume accelerates faster than that of the total villous surface area. The relative growth of the area is retarded in relation to the growth of the volume, so that the area/volume ratio is progressively reduced. This can also be expressed by stating that the available relative villous surface area (S_R) is larger in the early development stages than it is towards the end of gestation.

The size of the total villous surface area corresponding to a given total volume can be read directly off the graphs (Figs. 18–23). This reveals differences between the various species studied. In the cow, for example, a total volume of 35000 cm³ (written for simplicity as 35 kg) corresponds to a total villous surface area of about 110 m² (Fig. 20) whereas in the horse the same volume corresponds to an area of only about 12 m² (Fig. 19). If we were to plot the data points for the placenta of the horse on the same diagram as those for the placenta of the cow, the data points for the horse would lie lower on the diagram than those for the cow, over the entire development age range studied. Such differences, however, stand out more clearly in logarithmic coordinates and will therefore be discussed in greater detail in connection with Figs. 25–27.

3.2. Transformed Values

In discussing the transformed values of the total villous surface area and of the total volume in Sections 2.1 and 2.2 above, we noted that these transformed values, i. e., the linear dimensions of the area and volume respectively, grow at constant rates. We may now ask whether, for each of the species studied, these two parameters grow at the same constant rate or whether one of the two grows more rapidly than the other. One way of checking this is by comparing the regression coefficients b shown in Column 2 of Tables 1 and 2, respectively. It can be seen that in all the species studied the coefficient b for the square root of the total villous surface area (Table 1) is larger than the coefficient b for the cube root of the total volume (Table 2), i. e., that the length of side of the square equivalent to the total villous surface area grows faster than the length of edge of the cube equivalent to the total volume. In cows, for example, the length of side of the area square increases by about 35 cm per week (Table 1, Column 2) whereas the length of edge of the volume cube increases by only about 1 cm per week (Table 2, Column 2).

Another way of bringing out these differences is by plotting the square root of the total villous surface area as ordinates against the cube root of the total volume as abscissae, in linear coordinates. As each of these two parameters grows at a constant rate, the resulting families of data points can be fitted with straight smoothing lines, one for each species (Fig. 24). Calculating these smoothing lines as regression straight lines, we

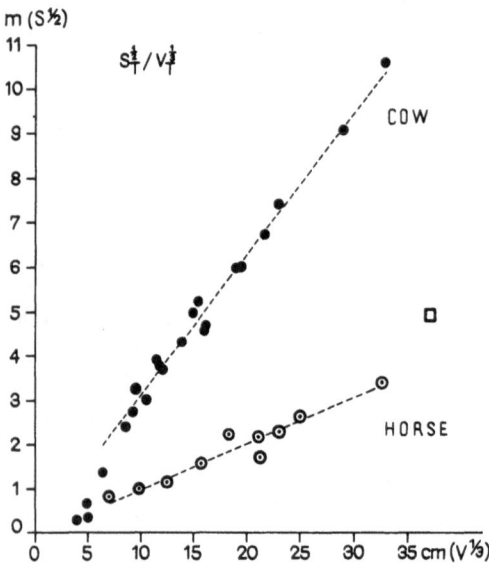

Fig. 24. Square root values of the total villous surface area (S_T) plotted against the cube root values of the total volume (V_T). Comparison between cow (full circles) and horse (dotted circles). In this mode of presentation, the growth rate of the linear dimension of the area (length of side of the equivalent square) is compared to that of the linear dimensions of the volume (length of edge of the equivalent cube). The square symbol represents the mean value of measurements carried out on 5 fresh full term placentas of the horse

find for all the species studied regression coefficients b > 1 (Table 4, Column 2), expressing the fact that the length of side of the area square increases more rapidly than the length of edge of the volume cube. These data also show, however, the extent of the differences in these relative rates of growth. In the cow, for example, the rate of growth of the length of side of the area square is greater than that of the length of edge of the volume cube by a factor of about 32, whereas in the horse this factor is only about 10 (Table 4, Column 2). Accordingly, the slope of the straight smoothing line is steeper for the cow than for the horse (Fig. 24). This observed difference between the cow and the horse may be considered statistically significant because the 0.05 confidence ranges of the two regression coefficients do not overlap (Table 4, Column 3). The ratios for the other species studied are also shown in Table 4. The lowest value was found for the pig.

Table 4. Square root of the total villous surface area (cm) versus the cube root of the total volume (cm)

	1 N	2 b	3 Conf.b (0.05)	5 a	6 r
Cow	17	31.93	29.9–34.0	– 8.6	0.993
Horse	10	10.50	8.3–12.7	– 8.7	0.969
Man[a]	38	21.60	20.5–22.7	–10.0	0.990
Cat[b]	17	15.61	12.9–18.3	9.7	0.954
Pig	18	7.27	6.4– 8.2	6.3	0.972
Rat	10	11.33	10.2–12.4	– 1.9	0.993

[a]Including data of Aherne and Dunnill (1966)
[b]Linear regression as a simplification

3.3. Logarithmic Coordinates

If we plot the total villous surface area (S_T) as ordinates against the total volume (V_T) as abscissae in logarithmic coordinates (with both axes on a logarithmic scale), we obtain for each species a family of points rising towards the right (Figs. 25–27). For

Figs. 25–27. Total villous surface area (S_T) plotted against the total volume (V_T) in logarithmic coordinates (both axes logarithmic). The available data points can be fitted with straight smoothing lines (except for the earliest stages of the cow). The dashed lines are the 95 % confidence range bands of the regression straight lines. In the case of the pig and the horse the macroscopic chorionic surface areas (S_M) are also shown (black triangles).

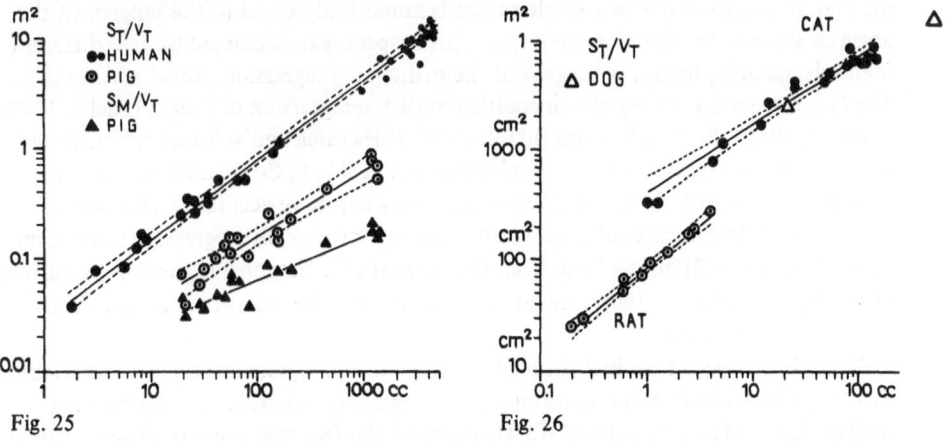

Fig. 25   Fig. 26

Fig. 25. Man (full circles) and pig (dotted circles and black triangles). In the case of man the data points on the left originate from our own measurements, those on the right (smaller dots) were calculated from the data of Aherne and Dunnill (1966). The total villous surface area of the human placenta is throughout larger than that of the pig placenta for comparable total volumes

Fig. 26. Cat (full circles) and rat (dotted circles). The two triangles show two data points for the German sheep-dog (mean values of several placentas and foetuses at the same development stages)

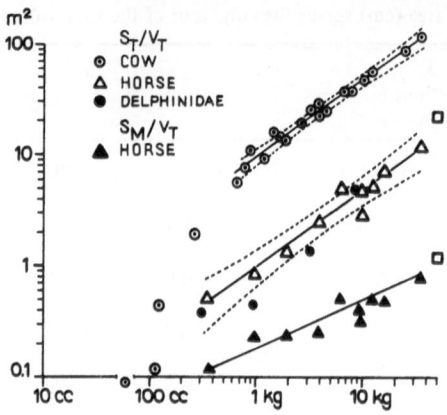

Fig. 27. Cow (dottet circles) and horse (white triangles for S_T, black triangles for S_M). The total villous surface area of the cow placenta is throughout larger than that of the horse placenta for comparable total volumes. The two squares show mean values of 5 fresh full term placentas of the horse (top square: S_T; bottom square: S_M). The four full circles show S_T values of placentas of delphinides

most of the species studied, the available measured data can be fitted with sufficient accuracy with a straight smoothing line. The data for the placenta of the cow, however show that such a straight-line fit is an approximation and cannot be used for the earliest development stages (Fig. 27).

Slope of the Straight Smoothing Lines

In considering the representation of the area/volume ratio in linear coordinates in section 3.1 above, we have already noted that the total villous surface area grows with a lesser acceleration than the total volume, and is therefore gradually retarded in relative terms (Figs. 18–23). The ratio of the accelerations of the rates of growth of the area and of the volume can be expressed numerically with the aid of plots in logarithmic coordinates. In those cases where the data points can be fitted with a straight smoothing line, this ratio of the two accelerations is numerically equal to the tangent of the angle of slope of the fitted straight line. This tangent was calculated by two different methods, namely, firstly, as tangent of the orthogonal regression (Sokal and Rohlf, 1969) and secondly, as regression coefficient of the regression of y on x (Sachs, 1972; Berkson, 1950). Bearing in mind that both variables (area and volume) are burdened with errors, the tangent of orthogonal regression (Table 5, Column 8) describes the general trend of each family of data points somewhat more accurately (Sokal and Rohlf, 1969). In the particular case of this study, however, the regression coefficient (Table 5, Column 2) differs little from the tangent of orthogonal regression (error not exceeding 2 %) and was therefore given preference for the sake of easier statistical assessment of the data.

With all the species studied the angle of slope of the regression straight line is smaller than 45° (Figs. 25–27) and, accordingly, the regression coefficient is significantly smaller than 1 (Table 5, Column 4). This reflects the fact that the rate of acceleration of the growth of the total villous surface area is smaller than that of the total volume.

The regression straight lines of the various species studied are roughly parallel (Figs. 25–27). Accordingly, the regression coefficients are also of roughly the same order, i. e., they vary only from about 0.6 to about 0.8 (Table 5, Column 2). The 0.01 confidence ranges of the regression coefficients of all the six species studied in detail overlap to a greater or lesser extent, i. e., no differences significant at the 0.01 level can be

28

Table 5. Log of the total villous surface area versus log of the total volume

	1	2	3	4	5	6	8
	N	b	Conf.b (0.1)	Conf.b (0.01)	a	r	tg
Cow[b]	17	0.672	0.624–0.720	0.592–0.753	938.0	0.988	0.677
Horse	10	0.690	0.589–0.791	0.509–0.872	80.4	0.976	0.701
Man[a]	38	0.716	0.698–0.733	0.688–0.743	290.4	0.996	0.717
Cat	17	0.611	0.548–0.674	0.504–0.718	420.1	0.975	0.620
Pig	18	0.598	0.529–0.666	0.483–0.712	102.3	0.967	0.610
Rat	10	0.784	0.731–0.838	0.689–0.880	86.7	0.995	0.788
Compact full term placentas	18	0.876	0.807–0.945	0.760–0.991	97.8	0.984	0.890
Diffuse full term placentas	11	0.883	0.657–1.109	0.482–1.284	12.6	0.922	0.954

[a] Including data of Aherne and Dunnill (1966)
[b] Earliest stages not included

detected between these six species on the basis of the data available (Table 5, Column 4). Some differences are found only with the use of the narrower 0.1 confidence ranges (Table 5, Column 3). The arithmetic mean value of the available six regression coefficients amounts to 0.679 or roughly 2/3. It may therefore be stated as a rough approximation that the ratio of the accelerations of the area and volume growth rates is about 2 : 3.

Size of the Total Villous Surface Area

From the data presented graphically in Figs. 25–27 we can read, for a given total volume, the total villous surface area available for the metabolic exchange between mother and foetus. In the human placenta, for example, the total villous surface area corresponding to a total volume of 1000 cm³ is about 4 m² (Fig. 25). Comparisons of the regression coefficients (Table 5, Columns 2–4) show that the straight smoothing lines for the various species studied may be considered as very nearly parallel. This makes it possible to compare with each other, for the various species studied, the sizes of the total villous surface area corresponding to the same total volumes.

If we compare the data points for diffuse and for compact placentas entered in the same coordinate systems, it is immediately obvious that the data points for diffuse placentas are located lower on the diagrams than those for compact placentas. This can be seen in Fig. 25, showing the data for the human placenta (compact placenta) and for the placenta of the pig (diffuse placenta), and in Fig. 27 which shows the data for the cow placenta (compact placenta) and for the horse placenta (diffuse placenta). These differences may be considered statistically significant because the corresponding 0.05 confidence range bands do not overlap in the domain studied.

If we compare all the species studied to each other, we find the following series: According to our measurements, by far the largest values of the total villous surface area for a given total volume are found in the placenta of the cow (during the second half of the gestation period, Fig. 27). The values for man (Fig. 25) and for the cat (Fig. 26) are smaller than those for the cow. If we were to plot the data for man and for the cat in the same coordinate system we would find that the values within the comparable range (1−100 cm³ total volume) are largely of the same order. For example, for a total volume of 100 cm³ the total villous surface area of the human placenta is about 0.8 m² (Fig. 25) whereas that of the cat placenta is about 0.7 m² (Fig. 26). The confidence range bands overlap in the comparable total volume range, so that no statistically significant difference is found between the data for man and those for the cat.

The values for the rat placenta (Fig. 26) are somewhat smaller than those for the placentas mentioned above. Finally, the smallest total villous surface areas for a given total volume are found in the diffuse placentas of the pig and of the horse (Figs. 25 and 27). No statistically significant difference is found between the pig and the horse, because the confidence range bands overlap in the comparable total volume range. For example, for a total volume of 1000 cm³ the width of the 0.05 confidence range band is 0.46−0.85 m² for the pig, and 0.65−1.40 m² for the horse.

Fig. 26 also shows two data points for placentas of the German sheep-dog (each triangle corresponds to the mean value of several foetuses and placentas of the same uterus). The placenta of the dog is of a structural type very similar to that of the cat. As can be seen in Fig. 26, these two data points are of the same order as those for the cat, and indicate the same growth trend. Finally, Fig. 27 shows, in addition to other data, four data points for delphinides (diffuse placenta, similar to that of the horse). These data points are in good agreement with those for the horse, in respect of the order of magnitude and of the indicated growth trend.

4. Relative Villous Surface Area

The relative villous surface area (S_R) is the area available to serve the metabolic exchange needs of 1 cm³ of tissue. It is another way of presenting the same data as those already presented in Figs. 18−23 and 25−27 in which the total villous surface area (S_T) is plotted against the total volume (V_T). It is therefore sufficient to present here only a few examples.

If we plot the relative villous surface area as ordinates against the total volume as abscissae in linear coordinates, we obtain a family of data points dropping to the rigth and which can be fitted with a hyperbolic smoothing curve (example: Fig. 34, p. 38). This shows that the relative villous surface area decreases in the course of development, relatively rapidly at first and then progressively more slowly. As a result, the villous surface area available for metabolic exchange per unit volume is smaller towards the end of gestation than in the earlier stages, i. e., the area/volume ratio becomes progressively less favourable in the course of development. This corresponds to the findings already presented in Figs. 18–23 and applies to all six species studied (for the cow, however, this applies only for the second half of the gestation period, see Baur, 1972).

In logarithmic coordinates the family of data points can usually be fitted with a straight line (example: Fig. 36, man), in accordance with the data shown in Figs. 25–27 where the total villous surface area is plotted against the total volume. This means that, within the range studied, the development of the relative villous surface area can be represented by a power function with, however, a negative exponent (straight line dropping towards the right).

The numerical value of this exponent, expressed by the regression coefficient of the straight smoothing line, varies in the different species studied from about −0.2 to about −0.4 (Table 6, Column 2). The confidence ranges of the various regression coef-

Table 6. Log of the relative villous surface area versus log of the total volume

	1 N	2 b	3 Conf.b (0.1)	5 a	6 r
Cow	17	−0.328	−0.376/−0.280	939.2	−0.952
Horse	10	−0.310	−0.411/−0.209	80.4	−0.897
Man[a]	38	−0.285	−0.302/−0.268	291.0	−0.978
Cat	17	−0.393	−0.457/−0.329	426.8	−0.941
Pig	18	−0.403	−0.471/−0.334	102.4	−0.932
Rat	10	−0.216	−0.269/−0.163	86.7	−0.937
Compact full term placentas	18	−0.125	−0.056/−0.193	97.8	−0.619
Diffuse full term placentas	12	−0.117	−0.329/ 0.095	12.6	−0.302[b]

[a] Including data of Aherne and Dunnill (1966)
[b] Not significant

ficients overlap (Table 6, Column 3), as do those shown in Table 5, i. e., the straight smoothing lines may be considered very nearly parallel. For the cow, this again applies only for the second half of the gestation period (Baur, 1972), whereas for the cat the family of data points shows a drooping (curved) trend (Fig. 35).

5. Volume of the Compact Parts of the Placenta

In compact placentas the growth of the total villous surface area is affected by two factors, namely, firstly, by the growth of the compact parts of the placenta (placentomes, placental girdle, placental disc) which contain the surface-enlarging structures and, secondly, by the extent of surface enlargement (surface density S_V) as a result of the internal differentiation of the compact parts of the placenta. In this section we shall consider the growth of the compact parts of the placenta without distinguishing between the maternal and the foetal components.

If we plot in linear coordinates the volume of the compact parts of the placenta of the cow (V_C, placentomes) as ordinates against the development age as abscissae, we obtained a family of data points which can be fitted with a curve concave upwards (Fig. 11). This shows that growth takes place at an increasing rate (acceleration). On the other hand, the same data expressed in terms of the cube root of the volume, and plotted in linear coordinates as ordinates against the development age as abscissae, result in a family of data points which can be fitted with a straight line (triangles in Fig. 12).

The cube root transformation amounts to considering a cube of the same volume as the volume of the compact parts of the placenta, and studying the growth of the linear dimension of this volume in the form of the length of edge of this equivalent cube. The straight smoothing line in Fig. 12 means that, in the cow, the linear dimension of the placentome volume (V_C), as represented by the length of edge of the equivalent cube, grows at a uniform, i. e., at a constant rate, at least in the development age range covered by the available data. This is a similar growth principle to that found for the total volume (V_T) the cube root values of which can also be fitted with a straight smoothing line (circles in Fig. 12).

With the placenta of the rat, the cube root values of the volume of the compact parts of the placenta (labyrinth) can also be fitted with a straight smoothing line (Fig. 16). With the human placenta (Fig. 15), however, and even more clearly with the placenta of the cat (Fig. 17), a straight line can be fitted only to the data points corresponding to the later stages of gestation, whereas the data points corresponding to the earlier stages show a steeper rising trend. In terms of the equivalent cube model this means that the length of edge of the equivalent cube grows rapidly at first, and that this rate of growth later slows down and becomes constant. The fact that, for humans and for cats, a straight smoothing line is not applicable to the entire gestation period is also evidenced by the calculation of the regression straight lines which, in the cases of man and of the cat, have positive y-axis intercepts a (Table 3, Column 5) i. e., if extended to the left, would intersect the x-axis (age) in the negative domain, which is biologically meaningless.

Regardless of the shape of the smoothing curve, however, all four compact placentas studied present the common feature that the volume of the compact parts of the placenta increases in the course of development (rising trends of the families of data points in Figs. 11, 12, 15, 16 and 17).

The growth of the compact parts of the placenta is of course slower than that of the total volume, as evidenced in Figs. 12, 15, 16 and 17 by the less steep rise of the corresponding families of data points (triangles versus circles).

6. Surface Density

The surface density (S_V) represents the villous surface area contained in 1 cm³ of the compact parts of the placenta (expressed in cm²/cm³). The data points plotted against the development age for rat, cat and man (Figs. 29—31) show a trend rising towards the right and can be fitted with straight smoothing lines. The calculations of the smoothing lines as regression straight lines show that the slopes are significantly different from zero (Table 7, Column 4).

Figs. 28—31. Development of the surface density (S_V). This is the average villous surface area (in cm²) contained in 1 cm³ of placental tissue, and is a measure of the enlargement of the exchange surface area caused by the villi in compact placentas. Except for the cow, the placentas of all the species studied show a marked increase in surface density in the course of gestation.

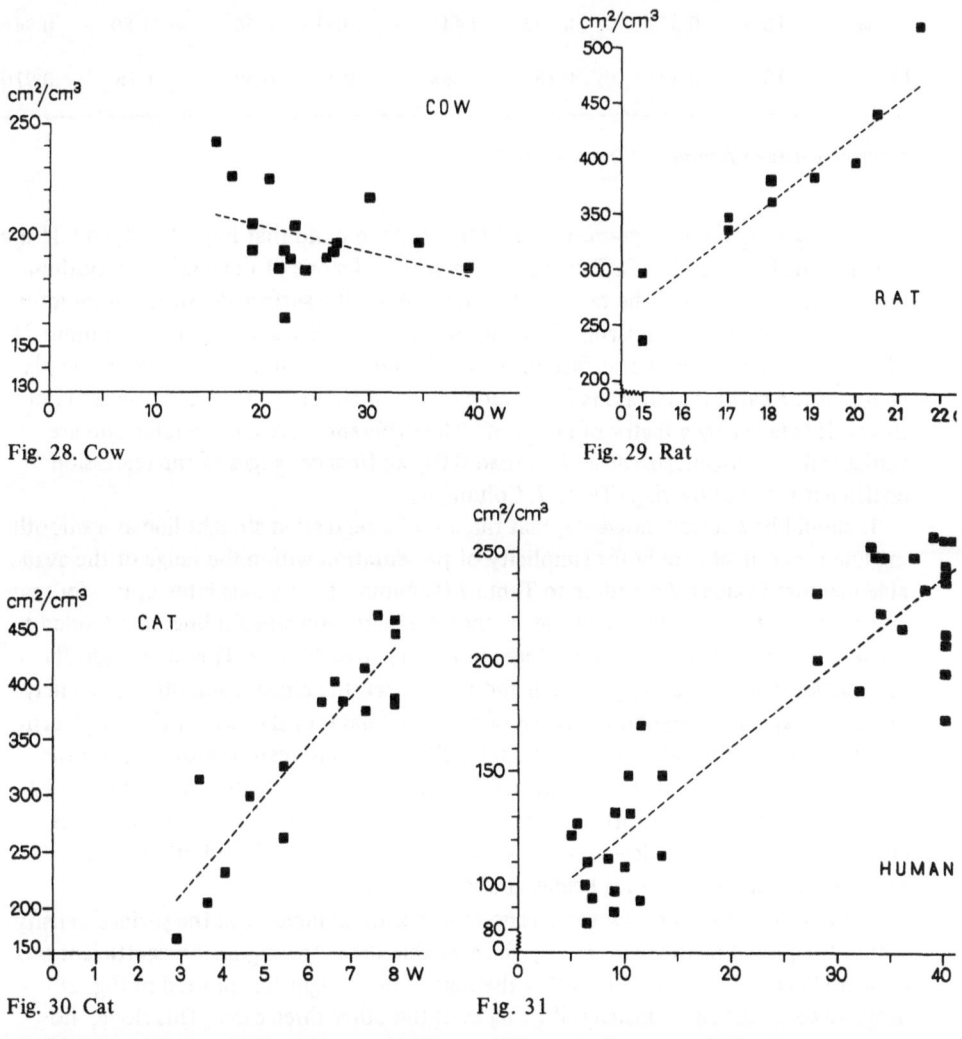

Fig. 28. Cow

Fig. 29. Rat

Fig. 30. Cat

Fig. 31

Fig. 31. Man. Data points on the left according to our own measurements, data points on the right according to the data of Aherne and Dunnill (1966)

33

Table 7. Surface density (cow, man, cat, rat) and surface enlargement factor (horse, pig), versus the development age (weeks)

	1 N	2 b	3 Conf.b (0.1)		4 Conf.b (0.01)		5 a	6 r
Cow	17	− 1.15	− 2.45 /	0.14	− 3.33 /	1.03	227.2	−0.373
Man[a]	38	3.88	3.38 −	4.36	3.06−	4.68	83.8	0.910
Cat	17	43.89	34.3 −	53.5	27.7 −	60.0	80.9	0.900
Rat	10	204.79	165.6	−244.0	134.2	−275.4	−164.7	−0.960
Horse	10	0.354	0.273−	0.44	0.21−	0.50	− 1.80	0.944
Pig	18	0.223	0.18 −	0.268	0.15−	0.30	1.16	0.910

[a]Including data of Aherne and Dunnill (1966)

In comparing these diagrams it should be borne in mind that Figs. 29, 30 and 31 are drawn to different scales. Differences in the angles of slope of the straight smoothing lines, i. e., differences in the rates of development of the surface density, can therefore be assessed better by comparing the linear regression coefficients (Table 7, Column 2). The slowest increase in the surface density is found in the human placenta; in the placenta of the cat this increase is faster by a factor of about 10, and in the placenta of the rat it is faster by a factor of about 50. The differences are considerable and are statistically significant, because the broad 0.01 confidence ranges of the regression coefficients do not overlap (Table 7, Column 4).

It should be stressed, however, that the use of a regression straight line as a smoothing line is permissible only for simplicity of presentation within the range of the available measured values. According to Table 7 (Column 5) the y-axis intercepts a for man and for the cat are positive. This means that the regression straight line, if extended to the left, would intersect the x-axis (age) in the negative domain. This is biologically meaningless and shows that, for man and for the cat, the straight smoothing line may not be extrapolated beyond the range of measured data. In the case of the rat placenta, however, this limitation does not apply. The y-axis intercept is negative, and the regression straight line intersects the x-axis at a point corresponding to about 11−12 days after mating. This information is valid because it is known from other studies (Franke, 1969) that the development of the placental labyrinth and of its surface-enlarging structures in the rat begins at about that time.

In the placenta of the cow there is no evidence of an increase in the surface density within the period of time covered by the available data. The regression coefficient b is negative (Table 7, Column 2), so that the regression straight line plotted in Fig. 28 drops towards the right, instead of rising as in the other three cases. This slope, however, is statistically not significantly different from zero, because even the narrow 0.1 confidence range straddles the value zero (Table 7, Column 3). Accordingly, the correlation coefficient r (Table 7, Column 6) is also not statistically significant, i. e., no correlation can be demonstrated between age and the surface density. In other words, the surface density of the placenta of the cow remains almost constant over the development period covered by the available data.

The above discussion relates to changes in the surface density during development. Another relevant point is the order of magnitude reached by the surface density in full term placentas of various species with compact placentas. Fig. 32 presents the frequen-

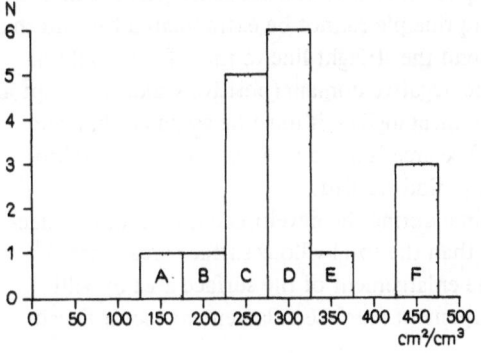

Fig. 32. Frequency distribution of the surface density (S_V) in full term placentas of 17 animal species with compact placentas. The distribution shows peaking in the classes of 250 and 300 cm²/cm³. A: Sloth; B: Cow; C: Man, chimpanzee, gorilla, German sheep-dog, giraffe; D: Macaca fascicularis, Colobus polycomos (guereza), Pygathrix nemea, seal, dwarf zebu, elephant; E: Sealion; F: Rat, cat, leopard

cy distribution of the surface density in full term placentas of several species. It can be seen that the surface densities range from about 100 to about 500 cm² per cm³ of compact placental tissue, with predominance of the density classes of 250 and 300 cm²/cm³.

The values for diffuse placentas, with villi sufficiently developed at full term to make the determination of a surface density meaningful, are of the same order as those for compact placentas. In the diffuse placentas of the horse, wild ass, zebra, llama, hippopotamus and aardvark (Orycteropus afer) we found values ranging from 125 to 325 cm²/cm³ (not included in the values on which Fig. 32 is based). Values in excess of 500 cm²/cm³ have been found so far only by Müller et al. (1967) in the placenta of the guinea-pig. It may be concluded from our findings that values in the range between 100 and 500 cm²/cm³ correspond best to the biological possibilities and requirements of enlargement of a surface by means of villi.

7. Macroscopic Chorionic Surface Area

In diffuse placentas the growth of the total villous surface area is affected by two factors, namely, by the growth of the macroscopically measurable surface area of the chorionic sac and by the further enlargement of this surface area by villi and folds. It can be seen from Figs. 1 and 4 that in the two placentas of the diffuse type included in this study, namely, pig and horse, the macroscopic chorionic surface area increases during the development period covered by the available data. In the horse, for example, the macroscopic chorionic surface area reaches about 1 m² at full term (the largest macroscopic chorionic surface area known so far, amounting to about 2 m², was found in the hippopotamus). Figs. 1 and 4, however, do not provide exact information on the type of growth curve.

Further information on the growth of the macroscopic chorionic surface area can be obtained by plotting the square root of this area against the development age (Figs. 6 and 7). This transformation amounts to considering a square with an area equal to the macroscopic chorionic surface area, and studying the growth of the length of side of this equivalent square during the development period. Figs. 6 and 7 show that, for the pig and the horse, the growth of the length of side of this square can be represented by a straight line. This means that, within the development period covered by the available data, the length of side of this square increases at a constant rate. It can also be seen, however, that this simple growth principle cannot be extrapolated beyond this period of time because, if we were to extend the straight line to the left, it would intersect the x-axis (development age) in the negative domain (positive y-axis intercept a, Table 3, Column 5). As this is biologically meaningless, it must be assumed that the growth of the macroscopic chorionic surface area is more rapid in the early development stages than during the development period studied.

It can also be seen that, at any given time during the development period, the macroscopic chorionic surface area is smaller than the total villous surface area (Figs. 1 and 4, and Figs. 6 and 7). This reflects the enlargement of the surface area by villi and folds. The extent of this surface enlargement (surface enlargement factor f) can be roughly estimated from these graphs.

If we plot the macroscopic chorionic surface area against the total volume in logarithmic coordinates, we obtain for each species (horse, pig) a family of data points which can be fitted with a straight smoothing line. The ratio of these two parameters can therefore be expressed by a power function. The exponent of this power function (logarithmic regression coefficient) amounts to 0.36 for the horse and to 0.40 for the pig. These two values are statistically not significantly different from each other, because the 0.1 confidence ranges overlap (0.28–0.45 for the pig, 0.34–0.46 for the horse).

In both cases the logarithmic regression coefficient of the macroscopic chorionic surface area (S_M), i. e., the exponent of the power function, is smaller than 1 and is also smaller than that for the corresponding total villous surface area $(S_T$, Table 5, Column 2). This means that, firstly, the growth of the macroscopic chorionic surface area lags behind that of the total volume in the course of development and, secondly, that this relative retardation is even more pronounced than that of the total villous surface area. It follows that, in order for the relative rate of growth of the total villous surface area to accelerate faster than that of the macroscopic chorionic surface area, the surface enlargement due to the villi (surface enlargement factor f) must increase in the course of development.

8. Surface Enlargement Factor

The surface enlargement factor f is a measure of the enlargement of the surface area of diffuse placentas due to the villi. If we plot this factor against the development age in linear coordinates, we obtain for each species (horse, pig) a family of data points rising towards the right (Fig. 33). This shows that the surface enlargement factor increases in the course of development, owing to the development of villi and folds. In the case of the horse the family of data points in the investigated range can be well represented by

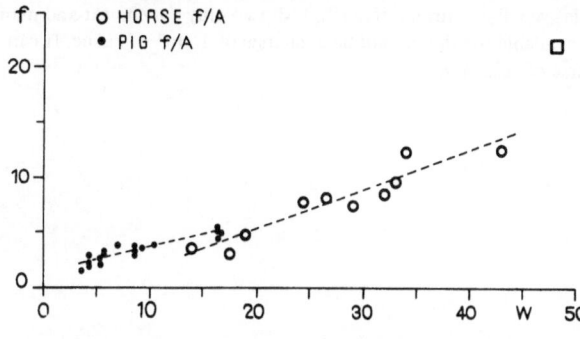

Fig. 33. Development of the surface enlargement factor (f). This is the factor by which the macroscopic surface area of the chorionic sac is enlarged in diffuse placentas by the presence of villi. dots: pig. circles: horse. The square shows the mean value of 5 fresh full term placentas of the horse

a straight line, indicating that the factor f increases at a constant rate. According to Table 7, Column 2, the rate of increase in the surface enlargement factor amounts to 0.354 per week.

In the case of the surface enlargement factor, extrapolations of the straight smoothing line towards earlier development stages are of course biologically meaningful only as far as $f = 1$ (values of f smaller than 1 would constitute a "diminution" factor). In the case of the horse, extrapolation of our findings shows that the value $f = 1$ corresponds to a development age of about 8 weeks, which corresponds approximately to the time when villi begin to develop. This finding supports the assumption of a straight smoothing line, i. e., of a uniform rate of increase in the surface enlargement factor.

With the placenta of the pig, a steeper slope of the smoothing line, i. e., a more rapid increase in the surface enlargement factor, must be assumed in the earlier development stages, because otherwise the value $f = 1$ would only be reached at negative values of x (development age), which is biologically meaningless.

It is also of interest to study the range of variation of the surface enlargement factor in full term diffuse placentas of various species (Table 8). The largest value found so far is 28.5 (Sardinian dwarf donkey). For the full term placenta of the pig, Table 8 shows two values of the surface enlargement factor, namely, 4.2 and 12.5, depending

Table 8. Surface enlargement factor (f) in full term placentas. – The data in brackets are the numbers of placentas studied. In the case of the pig the data shown are mean values of several placentas from the same uterus

Dwarf hippopotamus (Choeropsis liberiensis)	9.9 (5)
Zebra	10.4 (5)
Llama	12.7 (9)
Dolphin	15.5 (1)
Hippopotamus (H. amphibicus)	16.4 (1)
Indian rhinoceros (Rh. unicornis)	16.4 (5)
Aardvark (Orycteropus afer)	16.5 (1)
Bactrian camel	19.6 (2)
Somalian wild ass	20.9 (2)
Horse (thoroughbred)	22.0 (4)
Shetland pony	22.0 (7)
Sardinian dwarf donkey	28.5 (1)
Pig, measuring method A	4.2 (1)
Pig, measuring method B	12.5 (1)

Figs. 34–36. Development of the relative villous surface area (S_R), shown only for the cat and man as examples. This is the surface area available for the metabolic exchange of 1 cm³ of tissue. It can be seen that S_R decreases in the course of gestation.

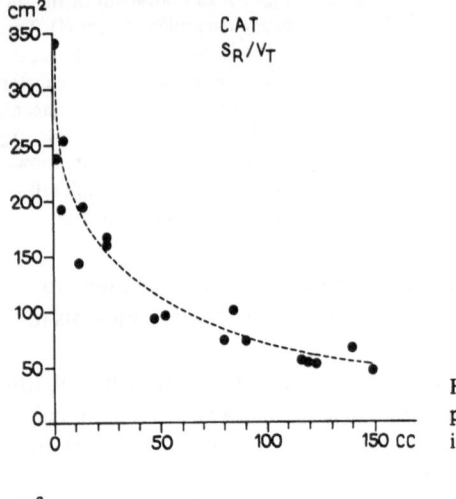

Fig. 34. Cat. In linear coordinates the available data points can best be fitted with a hyperboloid smoothing curve

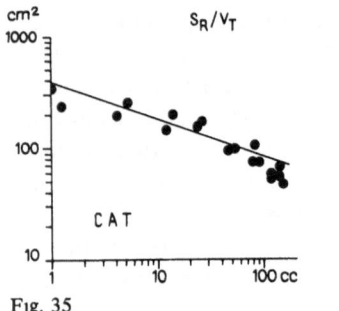

Fig. 35

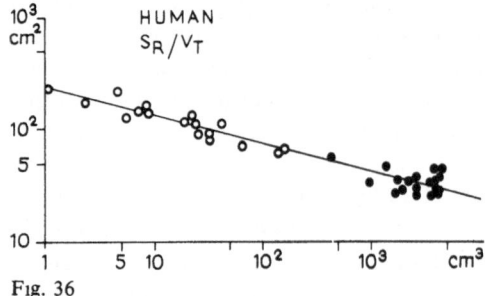

Fig. 36

Fig. 35. Cat. Same data as in Fig. 34, plotted in logarithmic coordinates

Fig. 36. Man. In logarithmic coordinates the family of data points can be fitted with a straight line dropping towards the right. Circles: our own measurements. Dots: values calculated from the data of Aherne and Dunnill (1966)

on whether the surface enlargement due to macroscopically visible folds is included for calculation purposes in the macroscopic or in the microscopic domain. The values for the other diffuse placentas range between 10 and 22 (with a mean value of 16.6, without the pig and the dwarf donkey). The general observation noted earlier in respect of the surface density of compact placentas seems also to apply to diffuse placentas, in that the surface enlargement factor of diffuse placentas of various species seems to vary within a common biological range.

9. Area/Volume Ratios in Full Term Placentas

In an earlier section we studied the variations of the inter-relationship of the total villous surface area (S_T) and the total volume (V_T) in the course of gestation in various species considered individually (intra-specific study of the growth trend). In this section we shall compare the relationship between the total villous surface area and the total volume in full term placentas of *different* species (inter-specific comparisons of analogous development stages). The data on compact and on diffuse placentas will be considered separately.

9.1. Compact Placentas

As can be seen from Fig. 37, in species with compact placentas included in this study the total volume (V_T) ranges over several orders of magnitude, namely, from 5.5 cm³ (rat) to 100 "kg" (elephant). The total villous surface area varies similarly, from 330 cm² (length of side of the equivalent square about 18 cm) in the full term placenta of the rat to about 300 m² (length of side of the equivalent square about 17 m) in the full term placenta of the elephant. This represents a range of 4 orders of magnitude.

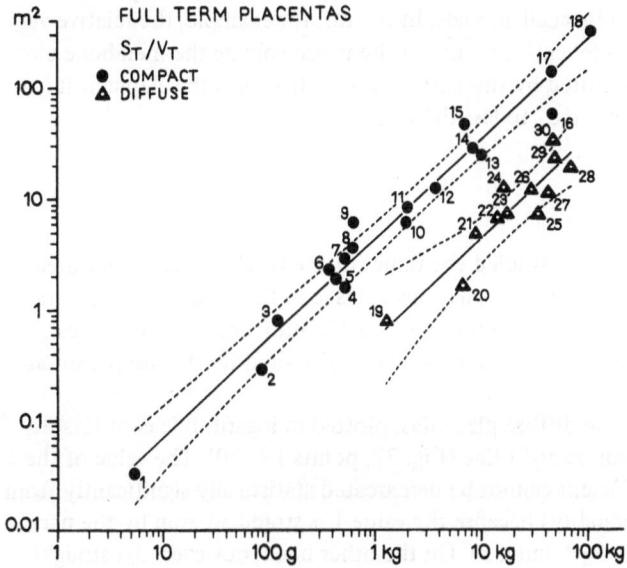

Fig. 37. Inter-specific comparisons. Full term placentas. Total villous surface areas (S_T) in various species, plotted in logarithmic coordinates against the corresponding total volumes (V_T). The families of data points for compact placentas (dots) and for diffuse placentas (triangles) were fitted with separate straight smoothing lines. The dashed lines show the 95 % confidence range bands of the regression straight lines. As these bands do not overlap within the range of the available data, it may be stated that diffuse placentas have significantly smaller total villous surface areas for comparable total volumes than the compact placentas.

Compact placentas: 1 Rat; 2 Guinea-pig (Cavia); 3 Cat; 4 German sheep-dog; 5 Sloth (choloepus didactylus); 6 Crab-eating monkey (Macaca fascicularis); 7 Douc langur (Pygathrix nemea); 8 Guereza (Colobus polycomos); 9 Leopard; 10 Chimpanzee; 11 Gorilla; 12 Man; 13 Dwarf zebu; 14 Seal (Phoca vitulina); 15 Sea-lion (Zalophus californianus); 16 Giraffe (Giraffa camelopardalis Tippelskirchi); 17 European domestic cow; 18 African elephant.

Diffuse placentas: 19 European domestic pig; 20 Dwarf hippopotamus (Choeropsis liberiensis); 21 Dolphin; 22 Llama; 23 Pony; 24 Sardinian dwarf donkey; 25 Zebra; 26 Somalian wild ass (Asinus asinus somalicus); 27 Bactrian camel; 28 Indian rhinoceros (Rh. unicornis); 29 Horse (thoroughbred); 30 Hippopotamus (H. amphibicus)

The data points 1—18 in Fig. 37 relate to species with compact placentas. It can be seen that when these data points are plotted in logarithmic coordinates (as in Fig. 37) their common trend can be represented by a straight smoothing line which may be considered as a regression straight line. Accordingly, the ratio of the total villous surface area to the total volume may be expressed by a power function as far as the full term placentas of these species are concerned. The straight line rises towards the right, i. e., the exponent of the power function is positive (positive value of the regression coefficient b in Table 5, Column 2). These data confirm the fact that, generally speaking, the placentas of large animals have larger total villous surface areas than those of small animals. On the other hand, the logarithmic regression coefficient of the area/volume ratios of compact placentas amounts to 0.876 (Table 5, Column 2) and is significantly smaller than 1 even at the 0.01 level (Table 5, Column 4). This means that the total villous surface area increases with the total tissue volume at less than a linearly proportional rate. The ratio of the total villous surface area (S_T) to the total volume (V_T) is therefore less favourable in animals with a large V_T than in those with a small V_T. In other words, the relative villous surface area (S_R) of the full term placenta is smaller in large animals than in small animals. In the rat, for example, the relative villous surface area amounts to 59 cm^2 per cm^3 of the tissue volume the metabolic exchange of which has to be ensured by this surface area, whereas in the elephant it amounts to only 26 cm^2/cm^3, i. e., to less than half.

9.2. Diffuse Placentas

In the diffuse placentas which we studied the values of the total villous surface area show a greater scatter than those in compact placentas, and the number of placentas studied is smaller. The findings are therefore less reliable statistically. Nevertheless, the findings described for compact placentas remain generally valid for diffuse placentas also.

The data points of full term diffuse placentas, plotted in logarithmic coordinates, can be fitted with a regression straight line (Fig. 37, points 19—30). The value of the logarithmic regression coefficient cannot be demarcated statistically significantly from the value 1 (linear proportionality) because the value 1 is straddled even by the narrow 0.1 confidence range (Table 5, Column 3). On the other hand, however, the straight smoothing line for the diffuse full term placentas is very nearly parallel to that for the compact full term placentas. This is confirmed by the very nearly equal regression coefficients, namely 0.883 for the diffuse and 0.876 for the compact placentas (Table 5, Column 2), and by the fact that there is no statistically significant difference between these two coefficients, because even the narrow 0.1 confidence ranges overlap to a considerable extent (Table 5, Column 3). It may therefore be stated that in diffuse full term placentas, as in compact full term placentas, the total villous surface area (S_T) is connected with the total tissue volume (V_T) by less than a linearly proportional rate, and that the area/volume ratio is consequently less favourable in larger animals.

9.3. Compact Versus Diffuse Placentas

Comparing the sizes of the total villous surface area (S_T) for the same or closely similar total volumes (V_T) in compact and in diffuse full term placentas, it is immediately obvious that the values for diffuse placentas are always lower than those for compact placentas over the entire range of species studied (Fig. 37). In the cow, for example, we find for a total volume (V_T) of 48 "kg" a total villous surface area (S_T) of 130 m² (compact placenta, No. 17 in Fig. 37) whereas for the horse we find for a closely comparable total volume (V_T) of 50 "kg" a total villous surface area (S_T) of only 22 m² (diffuse placenta, No. 29 in Fig. 37). We thus obtain for the full term placenta of the cow a relative villous surface area (S_R) of 27.0 cm²/cm³, whereas in the horse this parameter amounts to only 4.4 cm²/cm³. Moreover the 0.05 confidence range bands of the regression straight lines plotted in Fig. 37, do not intersect anywhere in the entire range of data studied. The differences between the relative villous surface areas (S_R) of compact and of diffuse placentas may therefore be considered statistically significant. It can be estimated from Fig. 37 that the relative villous surface areas of diffuse placentas are smaller than those of compact placentas by a factor of about 7.5 on average.

Full term diffuse placentas thus have significantly smaller villous surface areas available for metabolic exchange between mother and foetus per unit volume of foetal tissue than those found in compact placentas or, in other words, the relative villous surface areas of diffuse placentas are smaller. Moreover, according to our findings, the diffuse placentas of horses, pigs and delphinides have significantly smaller total villous surface areas than those of comparable compact placentas not only at full term but also *during* the gestation period, at least within the range of development stages studied (see Figs. 25 and 27). It may be assumed that this also applies to other diffuse placentas for which this difference has been proven only on full term specimens.

D. Discussion

1. Smoothing Curves and Function Equations

1.1. Growth of the Total Villous Surface Area

If we plot the total villous surface area (S_T) against the development age (A) in linear coordinates, we obtain for five of the six species studied, except the cat, families of data points rising towards the right and showing a trend concave upwards (Figs. 1–4), indicating an increase in the rate of growth with time. In the case of the placenta of the cat the available data points can be fitted with a straight line. In none of the species studied, however, do the measured data available so far indicate a slowing down of the growth rate towards the end of gestation, to say nothing of an actual arrest of placental growth, in contrast to the findings reported by Stegeman (1974) for the placenta of Texel sheep. It appears probable that the areal growth of the placentas which we studied is interrupted by parturition before any slowing down of the growth rate takes place.

In view of these findings we used in this study, as smoothing lines representing the growth of the total villous surface area, not sigmoid curves but sections of parabolas of the second degree (Figs. 1–4). This corresponds to a choice of power functions with the exponent 2 as the corresponding function equations. This choice also resulted from the study of the growth of the total villous surface area after square root transformation. In this transformation the total villous surface area is assimilated to a square with an equal area, and the variable studied is the length of side of this equivalent square (the linear dimension of the area). According to our findings, the families of data points obtained by plotting the square root of the total villous surface area against the development age, in linear coordinates, can be fitted with sufficient accuracy with straight smoothing lines for five out of the six species studied (except the cat) (Figs. 5–10). The growth of the linear dimensions of the total villous surface area can therefore be represented analytically by a linear equation of the type

$$y = a + bx \tag{1b}$$

where y is the length of side of the equivalent square in cm, x is the development age in weeks, a is the intercept on the y-axis, i. e., the value of y for x = 0, and b is the tangent of the angle of slope, i. e., the rate of growth of the linear dimension of the area. The parameters a and b for the species studied were determined by regression calculations and are shown in Table II, Columns 5 and 2.

Squaring both sides of Eq. (1b) and re-defining the variable y, we obtain for the growth of the total villous surface area a general equation of the form

$$y = (a + bx)^2 \tag{3}$$

42

where a, b and x have the same meanings as in Eq. (1b) whereas y now designates the total villous surface area in cm² (instead of its square root). Eq. (3) was used to calculate the smoothing curves shown in Figs. 1—4. As can be seen from these diagrams, these smoothing curves are a good fit for the available measured values. Until further data are available, however, it cannot be decided whether extrapolations beyond the range of measured data are permissible. Eq (3) is applicable for human placentas and for the placentas of the cow, horse and pig. In the case of the placenta of the rat the fit of the calculated curve is poor; the smoothing curve in Fig. 2 was therefore drawn free-hand and not according to Eq. (3). Our findings show that Eq. (3) is also not applicable for the placenta of the cat (Fig. 3).

In the case of the diffuse placentas of the horse and the pig, the growth of the macroscopic chorionic surface area (S_M) can also be represented by a straight smoothing line after square root transformation (Figs. 6—7). Eq. (3) can therefore also be used to represent the growth of the macroscopic chorionic surface area, but only within the development age range covered by the available measured data. This limitation arises from the fact that in both these cases (horse and pig) the straight smoothing line intersects the x-axis (development age) in the negative domain (positive values of the y-axis intercept a, Table 3, Column 5). This is biologically meaningless because it would mean that a chorionic surface already exists at the time of conception.

1.2. Growth of the Total Volume

Our findings show that, like the growth rate of the total villous surface area, the growth rate of the total volume (V_T) also accelerates up to parturition. According to Widdowson (1968) there takes place in humans a slowing down of the growth rate in the last phase of foetal development (after the 36th week), because of insufficient metabolic support. Also according to Widdowson this does not apply for rodents, carnivores and pigs, because the young of these species are born before the metabolic support becomes insufficient. According to Ostwald (1908), however, the phase of slowing down in the volumetric growth curve, as opposed to growth in length, begins with human foetuses also only after birth. This is confirmed by the curves plotted by Janisch (1927). in this study we have considered the growth of the total volume (V_T) in relation to the growth of the total villous surface area (S_T) only during the foetal development up to birth. According to our findings, the volumetric growth up to birth can be represented with sufficient accuracy by a continuously ascending curve. The choice of a sigmoid curve would be necessary only for following the volumetric growth after birth.

The choice of a suitable smoothing line results from the study of the growth of the total volume after cube root transformation. In this transformation the total volume is assimilated to a cube with an equal volume, and the variable studied is the length of edge of this equivalent cube (the linear dimension of the volume). According to our findings, the families of data points obtained by plotting the cube root of the total volume against the development age, in linear coordinates, can be fitted with sufficient accuracy with straight smoothing lines for all six species studied (Figs. 12—17). The same growth principle was found by Huggett and Widdas (1951) who studied the weight development of mammalian embryos and foetuses. On the basis of these findings the growth of the linear dimensions of the total volume can be represented analytically by a linear equation of the type

$$y = a + bx \qquad (1c)$$

where y is the length of edge of the equivalent cube in cm, x is the development age in weeks, a is the intercept on the y-axis, i. e., the value of y for x = 0, and b is the tangent of the angle of slope, i. e., the rate of growth of the linear dimension of the volume. The parameters a and b for the species studied are shown in Table 2, Columns 5 and 2.

Raising both sides of Eq. (1c) to the third power and re-defining the variable y, we obtain for the growth of the total volume a general equation of the form

$$y = (a + bx)^3 \qquad (4)$$

where a, b and x have the same meanings as in Eq. (1c) whereas y now designates the total volume in cm³ (instead of its cube root).

According to our findings, Eq. (4) is applicable for all six species studied, provided it remains limited to the development period covered by the measured data. Extrapolation into the post-natal time is not permissible.

1.3. Area/Volume Ratio during Development

If we plot for the various species studied the values of the total villous surface area (S_T) as ordinates against the corresponding values of the total volume (V_T) as abscissae, in linear coordinates, we obtain for each species a family of data points which can be represented by a smoothing curve convex upwards (Figs. 18−23). This means that the area/volume ratio is gradually reduced in the course of development.

The smoothing curves in Figs. 18−23 were calculated on the basis of the observation that, if we replace the area by its square root and the volume by its cube root, and plot these transformed values against each other in linear coordinates, we obtain for each species a family of data points which can be represented by a straight smoothing line (Fig. 24 and Table 4) which can be defined analytically by a linear equation of the type

$$y = a + bx \qquad (1d)$$

where y is the length of side of the equivalent square and x is the length of edge of the equivalent cube.

If we now re-define the variables y and x so that y represents the total villous surface area in cm² (instead of its square root) and x represents the total volume in cm³ (instead of its cube root), Eq. (1d) may be written in the form

$$y = (a + b \sqrt[3]{x})^2 \qquad (5)$$

This is the equation used for plotting the smoothing curves shown in Figs. 18−23, using in each case the parameters a and b shown in Table 4 for the species concerned. It can be seen from Figs. 18−23 that the fit achieved is satisfactory for all the species studied.

Instead of Eq. (5) it is also possible to use a power function of the general form

$$y = ax^b \qquad (2b)$$

where y is the total villous surface area in cm² and x is the total volume in cm³ . It

44

should be noted that the notations a and b in the above equation are used here in the statistical sense (see Sachs, 1972), i. e., in a different sense from that used in the allometric equation (Teissier, 1948) where they are reversed.

We shall discuss below the possible methods for determining the parameters a and b appearing in Eq. (2b).

The first method is based on the results of the root transformation. We have established that the linear dimensions of the area (length of side of the equivalent square) and of the volume (length of edge of the equivalent cube) plotted in linear coordinates against the development age (Figs. 5−10 and Figs. 12−17), as well as against each other (Fig. 24), can be represented by straight smoothing lines. The fact that these lines are straight indicates that the actually measured values tend to follow the theoretical geometric principle that a surface area increases as the square of its linear dimension and a volume increases as the cube of its linear dimension. It follows from these considerations that, theoretically, the exponent b in Eq. (2b) should have the value $b = 2/3$.

The parameter a in Eq. (2b) varies from species to species and depends on the ratio of the rates of growth of the linear dimensions of the area and of the volume, respectively. It can be obtained in theory from Table 4 by squaring the value of the regression coefficient (Table 4, Column 2; an even more accurate value would be obtained by using the tangent of orthogonal regression).

Another possible method for determining the parameters a and b of Eq. (2b) is based on the observation that, if we plot the total villous surface area against the total volume in logarithmic coordinates, the families of data points can be represented by straight smoothing lines (Figs. 25−27) which can be calculated as regression straight lines. The parameter a then corresponds to the y-axis intercept at $x = 1$ (Table 5, Column 5) whereas the parameter b corresponds to the regression coefficient of logarithmic regression (Table 5, Column 2; an even more accurate value would be the tangent of orthogonal regression, Table 5, Column 8). In this manner it is possible to establish an equation expressing the total villous surface area corresponding to a given total volume.

For example, applying Eq. (2b) to the placenta of the horse and using the values of a and b given in Table 5, Columns 5 and 2, we obtain

$$y = 80.4 \cdot x^{0.69} \tag{6}$$

where y is the total villous surface area in cm^2 and x is the development age in weeks. This equation is valid only within the development range studied, up to parturition, and should not be extrapolated to earlier development stages. It contains no correction allowance for the effect of histological shrinkage.

The calculation of the exponent b in Eq. (2b) as regression coefficient of logarithmic regression (tangent of the angle of slope of the regression straight lines in Figs. 25−27) offers the possibility of comparing statistically the values of b based on observations with the hypothetical value $b = 2/3$ arrived at earlier. It can be seen from Table 5, Column 3, that for the placentas of the cow, horse, cat and pig the value $b = 2/3 = 0.667$ lies within the narrow 0.1 confidence ranges for these species. This shows that, for these four species, it is permissible to use for the exponent b in Eq. (2b) the value $b = 2/3$. In the cases of man and the rat, however, the value $b = 0.667$ lies outside even the broad 0.01 confidence range limits (Table 5, Column 4), which means that the calculated values of b (Table 5, Column 2) are significantly different from $b = 2/3$. This shows that although the hypothetical value b = 2/3 is a practically usable and theoretically informative approximation it cannot be considered as a generally valid "law".

In all the species studied the data point families representing the area/volume ratios in logarithmic coordinates rise towards the right (Figs. 25–27), and the regression coefficients are positive and are statistically significantly different from zero at the 0.01 level (Table 5, Column 4). Accordingly, the correlation coefficients are also positive and statistically different from zero (Table 5, Column 6).

This proves statistically, for the species studied, the existence of a correlation between the total villous surface area (S_T) and the corresponding total volume (V_T) in the sense that, as this volume increases, this area increases also. A positive correlation was also found by Stegeman (1974) between the total villous surface area and the foetal weight in sheep. The statistical findings provide no information on the causal basis of this correlation (see Koller, 1963), but it may be assumed that it is a functional correlation. Bearing in mind the statistically relatively small sizes of the data samples, the correlation coefficients show very high values (very nearly 1, see Table 5, Column 6). The correlation is therefore very strong. This means also that the scatter of the data points about the regression straight lines in Figs. 25–27 is small, and that the regression straight lines are a good fit for the measured data.

It should be stressed, however, that the straight smoothing lines for the area/volume ratio in logarithmic coordinates are valid for each species only within the development age range studied so far and may not be extrapolated beyond this range. For the cow, in particular, the straight smoothing line is valid only for the second half of the gestation period. It can be seen from Fig. 27 that, in the cow, the data points corresponding to the early stages of gestation follow a steeper course which then changes into the 2/3 slope (logarithmic) at a total volume (V_T) of about 1000 cm³ (1 kg). It is interesting to note in this connection that Abeloos (1965) found in allometric studies of calf foetuses a slowing down of the growth rates of various organs in foetuses weighing about 1 kg. It is possible that the growth ratios before and after the development of the typical placental structure may also differ in placentas of other species, possibly in connection with a change in the mode of nutrition, e. g., in man, in connection with the transition from histiotrophic to haemotrophic nutrition. The study of these very early stages, however, requires different measuring methods than those used in this study and must therefore be deferred to a later date.

1.4. Relative Villous Surface Area

The study of the relative villous surface area (S_R), i. e., of the villous surface area available for the metabolic exchange needs of 1 cm³ of foetal tissue, is based on a somewhat different way of presenting the same data as those used in the study of the surface/volume relationship (Table 5 and Figs. 18–23 and 25–27). This study therefore yields no important new data but it does make it possible to bring out more clearly certain peculiarities of the surface/volume relationship.

We have seen that the variation of the total villous surface area (S_T) in function of the total volume (V_T) served by this surface area can be expressed by means of a power function. The same applies to the variation of the relative villous surface (S_R) in function of the total volume (V_T), but the exponent of the power function is smaller by 1 because, to obtain this function, S_T has to be divided by V_T not once but twice. Therefore, if we assume as a simplified model for the total villous surface area a power function with the exponent $b = 2/3$ as established earlier, we obtain for the relative villous surface area a power function with an exponent $b = 2/3 - 1 = -1/3$. The model equation then becomes

$$y = ax^{-1/3} \tag{7}$$

where y is the relative villous surface area (S_R) and x is the corresponding total volume (V_T). The values of the parameter a for the relative villous surface area (Table 6, Column 5) are in principle the same as those for the total villous surface areas (Table 5, Column 5). In practice, the computed values may show slight differences. The statistical significance assessments are the same as for the total villous surface area, i. e., the hypothetical value $b = -1/3$ is compatible with the findings for the placentas of the cow, horse, cat and pig, but not with those for man and the rat (Table 6, Column 3). This shows that the hypothetical value $b = -1/3$ yields a good model for most of the species studied but cannot be considered as a generally valid law.

The exponent b of the power function for the relative villous surface is negative in all the cases studied. This corresponds in linear coordinates to a hyperbolically dropping smoothing curve (Fig. 34), and in logarithmic coordinates to a straight line dropping towards the right (Figs. 35 and 36). In the case of man (Fig. 36), the plot in logarithmic coordinates shows a good fit of the straight line to the measured data. The less good fit in the case of the cat (Fig. 35) again shows that the power function used is only an approximation and cannot be considered as a "law of growth".

Regardless of the choice of the function equation, it may be stated that in all the species studied the relative villous surface area decreases towards the end of the gestation period. This means that the villous surface area available per unit volume of foetal tissue is smaller towards the end of the development period than in the early development stages. The exact evolution of the relative villous surface area during the earliest development stages is not yet clear. If we were to extrapolate the power function (in this particular case, a hyperbola) towards the earliest stages, we would obtain for the beginning of the development a value of the relative villous surface area tending to infinity. This would mean that some surface area would exist at a time when there is yet no volume, which is biologically meaningless. In fact, studies on the cow have shown that in the earliest development stages the relative villous surface area is smaller, so that the following picture of the development emerges (Baur, 1972): At first an increase in the relative villous surface area up to a maximum (in the cow: about 125 cm²/ cm³ at about 20 weeks), then a phase of progressive decrease up to birth; during this latter phase the development can be described by a power function with the exponent $b = -1/3$.

It appears likely that the same general principle also applies for other mammals, but with different timings of the transition. In man, for example, the phase of progressive decrease in the relative villous surface area begins at an earlier development stage (see Fig. 36, and similarly for the cat, Fig. 35), so that the power function can be used over a longer development phase than in the case of the cow. In the cases of man and of the cat, the graphs in Figs. 36 and 35 show that the increase in the relative villous surface area in the earliest development stages must be very rapid, taking place before the total volume reaches 1 cm³.

1.5. Area/Volume Ratios in Full Term Placentas

We have seen (Fig. 37) that the area/volume ratios in full term placentas, plotted in logarithmic coordinates, can be expressed by straight lines (one each for compact and for diffuse placentas). Such relationships can be expressed analytically by power functions of the type of Eq. (2b). Using the parameter values shown in Table 5, Columns 2 and 5, we can thus establish formulas expressing the total villous surface area of a full term placenta for a given total volume (new-born animal plus placenta). We thus obtain

for compact placentas: $y = 97.8 \cdot x^{0.876}$ \hfill (8)

for diffuse placentas: $y = 12.6 \cdot x^{0.883}$ \hfill (9)

where y is the total villous surface area in cm² and x is the total volume in cm³.

When using formulas (8) and (9) it should be borne in mind that the total villous surface areas calculated in this manner include no allowances for shrinkage errors or for possible further enlargements of the surface area by structures in the sub-microscopic range.

2. Compact Parts of the Placenta and Surface Density

In compact placentas, the growth of the exchange surface area of the placenta (S_T) may be affected morphologically by two factors, namely, firstly, by the increase in volume of the villi-containing compact parts of the placenta (V_C, placentomes, etc.) and/or, secondly, by an increase in the fine differentiation of the existing compact parts of the placenta, i. e., by an increase in the surface density (S_V).

According to our findings, both these factors play a role in the placenta of the cat, rat and man. In all three species the volume of the compact parts of the placenta (placental disc of the human placenta, placental labyrinth in the placentas of the cat and the rat) increases in the course of development (Figs. 15–17). The surface density also shows a marked increase in the course of development in all three species (Figs. 29–31).

From the data points for the human placenta, shown in Fig. 31, it is obvious that the surface density increases in the course of gestation, but the exact time-course of this development is difficult to assess. The increase in the surface density of the human placenta is confirmed by the findings of Cattoor (1967), but he used as smoothing line for his data points an exponential curve instead of a straight line as we have used in Fig. 31. An increase in the surface density means morphologically a finer differentiation of the surface-enlarging structures. In the human placenta this takes the form of the villi becoming thinner (Knopp, 1960; Hörmann, 1951), but this effect is probably further enhanced by a more profuse branching of the villi. According to the findings of Fujikura et al. (1971) the number of villous sections counted in a visual field of a given size increases in the course of development. This finding indicates an increase in the degree of branching of the villi.

The data on the placenta of the cow are somewhat different. According to our findings, there is no increase in surface density in the placenta of the cow during the second half of gestation (Fig. 28). It follows that, in the cow, the increase in the total villous surface area during the second half of gestation is ensured only by the increase in volume of the compact parts of the placenta (placentomes) (Figs. 11–12). Accordingly, our findings also show that the rate of growth of the volume of the compact parts of the placenta is greater in the cow than in the other species with compact placentas which we studied (Table 3, Column 2). On the other hand it may be assumed that in the cow there must take place an increase in the surface density in the first half of gestation so that, considering the family of data points in Fig. 28 as a whole, the development trend would be represented by an initially ascending line which then changes into a very nearly horizontal line during the second half of gestation. It would also be of interest to establish, by means of an inter-specific comparison of full term placentas of different species, whether there exists a correlation between the size of an animal and the surface density of the placenta, for example, whether in placentas with a small total volume the necessary surface enlargement is achieved by finer differentiation of the compact parts of the placenta. Such comparisons, however, are only meaningful between placentas of the same structural type. The available data on comparable placentas are insufficient for assessing whether such a correlation exists or not.

3. Macroscopic Chorionic Surface Area and Surface Enlargement Factor

Our findings on the diffuse placentas of the horse and the pig have shown that the macroscopically measurable surface area of the chorionic sac is smaller than the total villous surface area and that it grows at a slower rate (Figs. 1, 4, 6, 7, 25, 27 and Table 3) This is due to the fact that the macroscopically measurable chorionic surface area is enlarged by the presence of villi and that the surface-enlarging effect of the villi increases in the course of development. The increase in the surface enlargement factor is confirmed by the data shown in Fig. 33.

In the placenta of the horse this increase is due to the villi getting longer and branching more profusely. In the placenta of the pig this increase is governed by two factors the interdependence of which is discussed in greater detail below.

In the placenta of the pig the surface enlargement of the chorion takes place in two stages, disregarding for the time being any additional enlargement in the electron-microscopic range. The first stage consists of macroscopically visible folds 3 to 5 mm high (occasionally up to 1 cm high) running transversely to the longitudinal axis of the chorionic sac. The second stage consists of small folds 0.15–0.35 mm high, visible under a magnifying glass, with short stump-like villi developing over some parts of their crests. This pattern leaves a free choice as to whether the surface enlargement due to the macroscopic folds should be counted as part of the macroscopic chorionic surface area or as part of the microscopic surface enlargement factor.

The surface enlargement factor attributable to the macroscopically visible folds amounts to about 3, that attributable to the microscopic folds and the villi amounts to about 4. The total surface enlargement factor ist thus about 12. This value is of the same order as the surface enlargement factors of the other diffuse placentas studied (Table 8). For reasons of comparability it is therefore preferable to include the surface enlargement due to the macroscopic folds in the microscopic surface enlargement factor (Table 8, pig, measuring method B).

On the other hand, when measuring the macroscopic chorionic surface area of the pig placenta it is technically simpler to spread out the placenta in such a manner that the macroscopic folds are smoothed out. In that case the surface areas of the macroscopic folds are included in the measured macroscopic chorionic surface area. This results in a relatively large macroscopic chorionic surface area (Figs. 1, 6 and 25) and a correspondingly smaller surface enlargement factor (Fig. 33 and Table 8, pig, measuring method A). It should therefore be stressed that these values do not reflect a special position of the pig placenta but are due simply to a particular method of measurement. This does not affect the total area of the available exchange surface.

In the placenta of the horse, the surface enlargement factor in full term placentas is higher than could have been estimated by extrapolating the data available for the development stages (Fig. 33). This is probably due to the fact that full term placentas were processed in the fresh condition whenever possible whereas the development stages had been preserved in formol for several years before they were studied.

4. Model of the Growth of the Total Villous Surface Area and of the Total Volume

According to our findings, the general correlation between the growth of the total villous surface area of the placenta (S_T) and the growth of the total volume of tissue (V_T) served by this surface area can be represented by a simple model. We imagine a cube the length of edge of which increases at a constant rate. With this mode of growth, the rate of growth of the square root of the surface area and the rate of growth of the cube root of the volume can be linked by a linear equation, in accordance with our findings with placentas. If a solid body increases in size in accordance with the principle of geometric similarity, i. e., retaining its proportions, its surface area increases as the square of its linear dimensions whereas its volume increases as the cube of these dimensions. The volumetric growth therefore has a higher rate of acceleration (third power) than the areal growth (second power), and the ratio of the acceleration rates (area versus volume) is 2/3.

Owing to the faster growth of the volume, the area/volume ratio decreases in the course of development, i. e., the relative surface area gradually becomes smaller. For example, if the length of edge of the cube is doubled, its volume increase by a factor of 8 whereas its surface area increases by a factor of only 4, so that the relative surface area decreases by a factor of 2. Taking a less extreme example and expressing it in percentage terms, if the volume of a cube increases by 10 % (i. e., by a factor of 1.1) its surface area increases by only 6.56 % (i. e., by a factor of 1.0656) so that its relative surface area is reduced.

With this mode of growth, if we plot the area against the volume in logarithmic coordinates we obtain a straight line the tangent of the angle of slope of wich amounts to 2/3. This is in good agreement with the families of data points of the placentas which we studied, plotted in the same manner (Figs. 25–27 and Table 5, Columns 2–4). It can further be seen from Figs. 18–23, plotted in linear coordinates, that the smoothing curves for the growth of the total villous surface area (S_T) and of the total volume (V_T), based on the cube model, are a good fit for the actually measured data. It may therefore be concluded from our findings that the model described above is a good aid for visualising the evolution of the total villous surface area of the placenta in function of the total volume of foetal tissue which it serves. The following limitations must, however, be borne in mind:

1. In the human placenta, and in the placenta of the rat, the logarithmic regression coefficients of the area/volume ratio, although they are relatively close to the hypothetical value 2/3, are nevertheless statistically significantly different from this value (Table 5, Column 4). This indicates that, besides the power ratio of 2:3 between the area and the volume, the growth of the total villous surface area must also be influenced by other mechanisms.

2. The model may be used only within the periods of time covered by this study and, in the case of the cow, only for the second half of the gestation period.

3. A fundamental difference between the placenta and the cube model is that, in the cube, the volume of the cube and the area of its enveloping surface are linked to each other geometrically, whereas in the placenta the relationship linking the total villous surface area and the total of tissue served by this area is not geometric but functional. In the case of the cube, the linear dimensions of the area and of the volume are identical (length of edge of the cube) and therefore grow at the same rate. In the case of the placenta, on the other hand, the linear dimensions of the volume increase less rapidly than those of the area (the regression coefficients b in Table 2 are smaller than those in Table 1). Moreover, there are also considerable inter-specific differences. In the cow, for example, the linear dimension (square root) of the total villous surface area increases faster than the linear dimension (cube root) of the corresponding total volume by a factor of about 32, whereas in the pig this factor is only about 7 (Table 4, Column 2).

4. Another difference between the cube model and actual placentas is in the value of the y-axis intercept a of logarithmic regression (Table 5, Column 5). This parameter defines the size of the surface area corresponding to a volume $x = 1$ (in this particular case, area in cm^2 corresponding to a volume of $1\ cm^3$). In the case of a cube, $a = 6$. The smallest possible value of a is that for a sphere, namely, $a = 4.836$. In actual placentas the values of this parameter are very much higher. In the cow, for example, $a = 938\ cm^2$ (Table 5, Column 5). This is about 200 times greater than the corresponding value of a sphere with the same volume. This finding reflects the fact that, unlike the cube model, the total exchange surface area of the placenta is spatially independent of the total tissue volume which it serves, and is enlarged by the villi.

5. Physiological Implications of the Area/Volume Ratio

5.1. Development Stages

If we express the ratio of the total villous surface area (S_T) to the total volume (V_T) by means of a power function, we find that in all the species studied the exponent of this function is significantly smaller than 1 (Table 5, Column 4). This means that the proportionality between the rate of growth of the surface and that of the volume is less than linear. As a result, the ratio of the total villous surface area to the total tissue volume which it serves becomes progressively less favourable, i. e., the relative villous surface area (S_R) descreases in the course of development (Fig. 34—36). This raises the question of how the metabolic exchange between mother and foetus can be ensured despite this reduction in the relative villous surface area. The morphological findings allow several possible explanations, but a decision in favour of one or another of these explanations cannot be reached without appropriate physiological investigations.

1. One possible explanation would be that the total villous surface area available during the early development stages is relatively too large, i. e., that it includes a reserve capacity which is only utilised during later stages as a result of the decrease in the relative villous surface area (i. e., owing to the strong increase in the total volume). This raises the further question of whether perhaps the placenta gradually becomes insufficient towards the end of gestation (see Widdowson, 1968).

2. Another possible explanation would be that the exchange efficiency of the placenta per unit area increases in the course of development, so that the necessary metabolic exchange remains ensured despite the decrease in the relative villous surface area. An indication that this might be the case is provided by the progressive increase in vascularisation, as demonstrated by Stegeman (1974) in the placenta of the sheep. According to Aherne (1975) the metabolic exchange efficiency of the human placenta increases in the course of development because, in later development stages, the villous blood vessels are located nearer to the periphery of the villi. Another factor acting in the same sense could be the reduction in layer thickness due to the disappearance of the cytotrophoblast and to the formation of thin layers of the syncytiotrophoblast underlain by capillaries (epithelial plaques, Amstutz, 1960).

3. A third possibility might be that the decrease in the relative villous surface area observed in the microscopic range is compensated by increasing surface enlargement in the sub-microscopic range (microvilli). This possibility has not been investigated in this study so far. This possibility is in any case excluded as far as the exchange area in the placental labyrinth of the rat and of the cat is concerned because, in both these species, hardly any microvilli can be found on the trophoblast in the placental labyrinth in full term placentas (Björkman, 1970; Franke, 1969).

4. A fourth possibility might be that the progressive reduction in the relative villous surface area is compensated by a progressive reduction in the relative metabolic rate of the foetus, i. e., in the metabolic rate per unit of body weight and time, in the course of development.

It has been shown in inter-specific comparisons of adult animals that large animals have lower energy requirements per unit of body weight than small animals (Kleiber, 1967 and 1947; Lehmann, 1951). The ratio of the energy requirements to the body weight can be expressed by a power function with an exponent of about 2/3 to 3/4 (Kleiber, 1967; Lehmann, 1951). This is of the same order as the exponent of the ratio of the total villous surface area to the total volume which we found with placentas (Table 5, Columns 2—4).

The question arises of whether this "principle of metabolic rate reduction" (Lehmann, 1951), demonstrated for adult animals, is also applicable to the intra-uterine phase of development. This would mean that older, i. e., larger foetuses of a given species have lower energy requirement per unit of tissue volume (or weight) and can therefore be adequately supported by a smaller relative villous surface area. According to the findings of Barcroft and Elsden (1946) and of Carlyle (1948), sheep foetuses do indeed show a reduction in the consumption of oxygen per unit of weight and time toward the end of intra-uterine development. This possibility is also considered by Meschia (1972). On the other hand, according to the findings of Dawes and of Assali (quoted in Villee, 1960), the consumption of oxygen by sheep foetuses, per unit of weight and of time, remains very nearly constant throughout the entire development period. Our morphometric findings point to the need for further studies of this question, preferably in several mammalian species.

If the principle of metabolic rate reduction is in fact applicable to the intra-uterine development phase, our findings would indicate that the relative metabolic requirements and the relative villous surface area decrease at about the same rate in the course of development. This would mean that the size of the total villous surface area remains optimally adapted to the changing metabolic requirements of the foetus, i. e., that it is neither larger nor smaller than required at any given time. These questions, however, cannot be clarified without appropriate physiological investigations.

5.2. Full Term Placentas

According to our findings, the ratio of the total villous surface area to the total volume can also be expressed by a power function in interspecific comparisons of full term placentas. With compact placentas the exponent of this power function is significantly smaller than 1 (Table 5, Column 4). This means that the proportionality between the total villous surface area and the total volume is less than linear, so that the relative villous surface are, i. e., the exchange area available per unit volume of foetal tissue, is smaller in large animals. In the full term placenta of the rat, for example, the relative villous surface area amounts to 59 cm^2/cm^3, whereas in that of the African elephant it amounts to only 26 cm^2/cm^3. This finding may possibly be linked to the fact, mentioned above, that large animal species have smaller relative energy requirements than small species (Kleiber, 1967; Lehmann, 1951). According to our findings, however, the exponent of the ratio of the total villous surface area to the total volume in compact full term placentas is significantly greater than the value b = 3/4 postulated by Kleiber (1967) for the exponent of the ratio of the energy requirements to the body weight (see Table 5, Column 4). This means that, besides adaptation to the "principle of metabolic rate reduction" (Lehmann, 1951), the phenomena studied must also be affected by other factors.

The situation with diffuse full term placentas appears to be similar. Plotted in logarithmic coordinates, the regression straight line for diffuse placentas is very nearly parallel to that for compact placentas (Fig. 37), the logarithmic regression coefficients being b = 0.883 for diffuse placentas and b = 0.876 for compact placentas (Table 5, Column 2). With diffuse placentas, however, the scatter of the data points about the logarithmic regression straight line is greater and the size of the available statistical sample is smaller, so that the logarithmic regression coefficient of the available data points cannot be demarcated significantly against either of the two hypothetical values 1 and 3/4.

6. Differences in the Size of the Villous Surface Area in Compact and in Diffuse Placentas

Our findings show that, for comparable total volumes (V_T), the total villous surface areas (S_T) of compact placentas are larger than those of diffuse placentas. In other words, the relative villous surface area (S_R), i. e., the area available for the metabolic exchange between mother and foetus per unit volume of foetal tissue, is smaller in diffuse placentas than in compact placentas. This is true both of full term placentas and of the development stages which we studied, at least within the period of time covered by the measured data (Fig. 37 and Figs. 25–27). As can be seen from Fig. 37, the difference between full term compact and diffuse placentas amounts to a factor of about 7.5. As the 0.05 confidence range bands of the two logarithmic regression straight lines do not overlap, this difference may be considered statistically significant. This raises the question of how adequate metabolic exchange between mother and foetus can be ensured in diffuse placentas despite the smaller available relative exchange surface area.

6.1. Effects of Shrinkage on the Macroscopic Data

The first question to be studied is whether the area size difference between the two types of placenta is perhaps an artefact due, for example, to shrinkage of the tissues. This might be the case, in particular, with the development stages of the horse placenta because these placentas were available for measurements only on the fixed condittion. In fact, the values measured in full term placentas, which were examined in the fresh condition, are higher than could have been expected on the basis of extrapolation of the data measured in the development stages (square symbols in Figs. 4, 7, 24 and 33). In the case of the horse, therefore, the values measured in full term placentas were not included in the calculations of the regression straight lines for the development stages, because these two sets of data were measured under different conditions. Extrapolating from the development stages with the aid of the data show in Table 4 and of Eq. (5), we obtain for the full term placenta of the horse a total villous surface area (S_T) of 14 m^2, whereas the value actually measured in fresh full term placentas amounts to 22m^2. The total villous surface areas of the development stages should therefore be multiplied by a correction factor of 1.57. This does not, however, eliminate the difference between compact and diffuse placentas, as can be seen from the data for full term placentas in Fig. 37 in which the data shown for the horse were measured on fresh full term placentas.

6.2. Effects of Shrinkage on the Microscopic Data

A correction for the effects of shrinkage due to the histological preparation procedure would affect all the placentas studied to an equal extent (except the placenta of the cow because, in the particular case of the cow, the data presented in this study have already been largely corrected for the effects of shrinkage, see Baur, 1973). Such a correction would therefore not affect the difference between compact and diffuse placentas. Moreover, earlier studies (Baur, 1973) have shown that even if the correction for shrinkage were applied only to diffuse placentas it would reduce but not eliminate the observed difference in the relative villous surface areas of compact and diffuse placentas, amounting to a factor of 7.5.

It must be stressed, however, that all the values presented in this study should be considered as minimum values and that the true villous surface areas are probably larger. Besides shrinkage, the measured size of the total villous surface area is also affected by other factors. For example, the findings of Gehr et al. (1974) show that, in the lungs, measurements of the alveolar surface area in the visual microscopic range, despite correction for shrinkage, result in lower values than those found by measurements in the electron-microscopic range. It may therefore be assumed that, if we apply concurrently the corrections resulting from the findings of Gehr et al. (1974) and of Baur (1973), the corrected villous surface areas would be two to three times as large as the uncorrected values presented in this study (without taking into account an additional surface enlargement by microvilli). Such a correction, however, would not affect the observed difference in the relative villous surface areas between compact and diffuse placentas, because the correction would affect placentas of both types equally.

It should also be mentioned that in the cases of the rat and the cat the values presented in this study are somewhat on the low side because the exchange surface area was measured only in the placental labyrinth, neglecting the exchange surface area in the peripheral region of the cat placenta (paraplacental area, Janssen, 1933) and in the vitelline placenta of the rat. Both these species, however, have compact placentas so that, if these additional areas were taken into account, the observed difference between compact and diffuse placentas would become even wider. Similarly, the exchange surface area of the cow placenta would be enlarged still further by taking into account the paraplacental area (intercotyledonary area), whereas with diffuse placentas the entire surface area is already taken into account.

6.3. Excess Surface Area

In looking for an explanation of the difference in the relative villous surface area between compact and diffuse placentas, we should also consider whether compact placentas perhaps have an excess surface area, i. e., a larger reserve capacity than that of diffuse placentas.

6.4. Duration of Development

Another possible explanation might be that in species with diffuse placentas the smaller relative villous surface area is compensated by a longer development time. Table 9 shows, however, that amongst the species studied there are cases in which species with diffuse placentas, despite a smaller villous surface area and despite a similar or even a larger total volume, have a *shorter* gestation time than comparable species with compact placentas.

The hypothesis of compensation of a smaller relative villous surface area by a longer development time is thus invalid at least for the cases shown in Table 9 unless, in the species with compact placentas used in these examples, long gestation times are simulated by delayed implantation (Starck, 1975).

Table 9. Relative villous surface areas (S_R) of some full term placentas and durations of gestation

Animal species	Type of placenta	S_R cm²/cm³	Duration of gestation (days)	S_T m²	V_T kg
Giraffe	Compact	11.5	446[a]–476[b]	53	46
Bactrian camel	Diffuse	2.6	389[c]	11	43
Macaca fascicularis	Compact	62.5	164[d]	2.5	0.4
Pig	Diffuse	5.8	109[e]–126[f]	0.75	1.3
Gorilla	Compact	42.5	251[g]–289[g]	8.5	2.0
Dwarf hippopotamus	Diffuse	2.4	200[a]	1.6	6.7

[a]Birth records of Basle Zoological Garden
[b]Naaktgeboren and Slijper, 1970
[c]Mehta et al., 1962
[d]Jewett and Dukelow, 1972
[e]Marrable, 1971
[f]Needham, 1963
[g]Kirchshofer, 1970

6.5. Microvilli

Yet another possible explanation could be that the additional enlargement of the villous surface area in the electron-microscopic range, owing to microvilli, is more pronounced in diffuse than in compact placentas.

The surface-enlarging structures of the placenta can be classified into three dimension ranges, namely, the macroscopic, the visual microscopic and the electron-microscopic. The determination of the surface area in each of these ranges presupposes knowledge of the surface area of the preceding range. This study is concerned only with the first two dimension ranges, i. e., the measured data presented in this study do not so far take into account the possibility of supplementary surface enlargement in the electron-microscopic range. We shall nevertheless briefly examine below the extent to which the results of the macroscopic and visual microscopic measurements might be altered by supplementary surface enlargement in the electron microscopic range, owing to microvilli, and the possible order of magnitude of this supplementary surface enlargement.

The surface enlargement factor of the microvilli, i. e., the factor by which the exchange surface area measured in the visual microscopic range is enlarged owing to the presence of microvilli, can be determined by means of the same procedure in the electron-microscopic range as that used to determine the surface enlargement factor in the visual microscopic range in diffuse placentas (Baur, 1973). Measurements carried out on electron-microphotographs of a full term placenta of the llama (the microphotographs were kindly made available by Mr. U. M. Spornitz of the Anatomical Institute of Basle University) yielded for the trophoblast areas containing microvilli the following values of the surface enlargement factor: mean: 7.53; maximum: 15.0; standard deviation: 3.27. Provisional measurements of electron-microphotographs published by other authors (e. g., Bjorkman, 1970) yielded the following values: horse: 10.8; pig: 8.4; cow: 6.0; man: mean 7.0, maximum 13.8. Björkman (1970) found for the pig a factor of 10. On the basis of these findings it may be assumed that, in full term placentas, the surface enlargement factor due to the microvilli is of the order of 6–10 (at most 15).

Dick et al. (1970) found in the oocytes of the toad, in the phase of maximum development of the microvilli, a surface enlargement factor of about 11, i. e., of the same order as that in placentas. According to Brown (1962), the microvillous surface enlargement factor in the epithelium of the small intestine ranges from 5–40, depending on whether the measurements relate to the tips or to the bases of the villi.

These numerical values, however, are maximum values which would be valid only if the entire surface area of the villi were occupied by microvilli. It is therefore not permissible to multiply the entire villous surface area determined by visual microscopy by the surface enlargement factor due to the microvilli.

It should be borne in mind that the trophoblast areas specialised for metabolic exchange by diffusion (epithelial plaques, see Ludwig, 1968) carry no brush-border structures but, at most, a few isolated stump-like microvilli. In the case of the human placenta, for example, Laga et al. (1973) assume that the presence of microvilli increases the total villous surface area from 16 m² to 75 m², i. e., by a factor of 4.7.

The difference in the sizes of the villous surface area, found between diffuse and compact placentas in the visual microscopic observation range, could be compensated only if the additional surface enlargement caused by microvilli were more pronounced in diffuse placentas than in compact placentas. According to our provisional findings, however, the microvillous surface enlargement factors in the investigated diffuse placentas of the llama, horse and pig are not much greater than those in the compact placentas of man and the cow. This does not apply, however, to the compact placentas of the rat and the cat. In these two species there is in the labyrinth of the full term placenta almost no surface enlargement by microvilli (see Björkman, 1970; Franke, 1969) so that, unlike the findings in the diffuse placentas, the values measured in the visual microscopic range are enlarged only insignificantly by taking into account the electron-microscopic findings.

It seems plausible to assume that the function of the microvilli is similar to that of the villi and consists of increasing the surface area available for metabolic exchange. It should be noted, however, that in the case of the microvilli this does not apply to all kinds of metabolic exchange. In fact, the presence of microvilli reduces the surface area available for that part of the metabolic exchange which takes place by pinocytosis, because pinocytosis takes place only over the areas of the cellular surface between the basis of the microvilli (see Fawcett, 1966, p. 416).

6.6. Physiological Differences

Another possible explanation of the differences in the size of the villous surface area in compact and in diffuse placentas could be that the smaller size of the villous surface area in diffuse placentas is compensated by a more intensive metabolic exchange activity. The factors playing a role in this connection might include differences in vascular morphology (see Moll, 1972).

The hypothesis of a more intensive metabolic exchange in diffuse placentas is supported in the case of the horse by the findings of Comline and Silver (1975). They found that the placenta of the horse is more efficient in respect of gas exchange than the placentas of ruminants which they studied (cow, sheep, goat). They found no such difference, however, between the diffuse placenta of the pig and the placentas of ruminants.

The functional significance of the differences in the size of the villous surface area in compact and in diffuse placentas thus remains unclear. It can be seen, however, that the morphometric findings may provide fresh stimulation for further physiological studies.

7. Morphometric Findings and Systematic Classification of Placentas

The morphometric findings can also play a role in the systematic classification of placentas. Of particular interest in this connection is the taxonomic position of the cotyledonary placentas, e. g., of the placenta of the cow. According to the criteria used by Portmann (1938), the cotyledonary placentas belong to the same group as the diffuse placentas, namely, to the group of "extended" placentas. According to the classification used by Ludwig (1968) several diffuse placentas, namely those of Artiodactyla (e. g., the placentas of the pig and the camel) are also included in the same group as the cotyledonary placentas, because of similarities in the histo-morphology of the villi. According to morphometric criteria, however, e. g., the relative villous surface area, our findings show a distinct demarcation between compact placentas (amongst which we also include the cotyledonary placentas) on the one hand, and the diffuse placentas (e. g., horse, pig, camel) on the other. As can be seen from this example, consideration of morphometric criteria may provide new points of view in the taxonomy of placentas.

E. Summary

1. This paper describes findings of morphometric investigations of placentas in the macroscopic and the visual microscopic ranges of observation. These investigations were concerned mainly with the growth of the exchange surface area (total villous surface area) of the placenta in relation to the total volume (tissue of the foetus plus placenta) served by this surface area.

2. The investigations were carried out on the development stages of placentas of six species, namely, cow, horse, man, pig, cat and rat, and on full term placentas of a total of 30 species.

3. In all the species studied a marked correlation was found between the growth of the total villous surface area and that of the total volume.

4. The total villous surface area and the total volume both grow at increasing rates, but the acceleration of the growth rate of the surface area is smaller than that of the volume.

5. As a result, the relative villous surface area decreases in the course of development, i. e., the exchange surface area available per unit volume is smaller for older foetuses than for younger foetuses of the same animal species.

6. As a simplification, it may be stated that the area/volume ratio can be represented over a large part of the total gestation period by a power function with the exponent 2/3. This principle of growth can be illustrated by the following numerical example: During a time period in which the total villous surface area increases by a factor of four, the total volume increases by a factor of eight, and the relative villous surface area decreases by a factor of two.

7. The findings mentioned under Points 3–6 above apply to the development stages of all the six species studied, but may not be extrapolated beyond the time periods studied.

8. The morphometric findings raise the question of how the metabolic needs of the developing foetus can be ensured despite the decreasing relative villous surface area. Amongst the various possible explanations, the two explanations which seem most likely are a reduction in the relative metabolic requirements of the foetus or an increase in the exchange intensity of the placenta per unit of surface area.

9. In the development stages studied, and even more so in the full term placentas, we found that placentas of the diffuse type (in which the villi are distributed over the entire chorion; examples: horse, pig) always have a smaller total villous surface area for a given total volume than placentas of the compact type (in which the villi are concentrated over particular areas of the chorion; example: man, cow).

10. This morphometric findings raises the question of how sufficient metabolic exchange can be ensured in diffuse placentas despite the smaller available exchange surface area. Amongst the various possible explanations, the most likely seems to be a

difference in the intensity of metabolic exchange per unit of surface area in placentas of different types.

11. Inter-specific comparisons of full term placentas show that larger animals have smaller relative villous surface areas. If we express the area/volume ratio by a power function we find that the exponent amounts to about 0.88 and that it is statistically significantly different from the value 2/3 mentioned above. This shows that the conditions governing inter-specific comparisons are different from those governing comparisons between small and large foetuses of the same species in the course of development (intraspecific comparisons).

12. The inter-specific comparisons of full term placentas raise the question of how the metabolic needs of the foetuses of large animals can be ensured despite smaller relative villous surface areas. The explanation may perhaps be linked with the "principle of metabolic rate reduction" (Lehmann, 1951) according to which larger animals have relatively smaller metabolic requirements.

Acknowledgement

The author wishes to express his thanks to the translation service of Hoffmann-Laroche Ltd. for the translation of the manuscript.

F. References

Abeloos, M.: La croissance et la régénération. In: Encyclopédie de la Pléiade. Vol. 18, Biologie, pp. 633, Parıs: Gallimard 1965

Aherne, W. A., Dunnıll, M. S.: Quantitatıve aspects of placental structure. J. Path. Bact. 91, 123–139 (1966)

Aherne, W.: Morphometry. In Gruenwald, P.: The placenta and its maternal supply line. Lancaster: MTP medıcal and technıcal publishing Co Ltd. 1975

Amstutz, E.: Beobachtungen uber die Reifung der Chorionzotten ın der menschlichen Placenta mit besonderer Berücksichtıgung der Epithelplatten. Acta anat. (Basel) 42, 12–30 (1960)

Barcroft, J., Elsden, S. R.: The oxygen consumptıon of the sheep fetus. J. Physiol. (London) 105, 25 P (1946)

Barka, T. and Anderson, P. J.: Hıstochemistry. New York-Evanston-London: Harper and Row 1965

Baur, R.: Quantitatıve Analyse des Wachstums der Zottenoberfläche bei der Placenta des Rindes und des Menschen. Z. Anat. Entwickl. Gesch. 136, 87–97 (1972)

Baur, R.: Notes on the use of Stereological methods in comparatıve placentology. Acta anat. (Basel) 86, Suppl. 1, 75–102 (1973)

Bender, H. G., Werner, C. H., Malzer, G.: Plazentaınsuffızienz. Med. Klin. 69, 1543–1547 (1974)

Berkson, J.: Are there two regressıons? J. Amer. statıst. Assoc. 45, 164–180 (1950)

Bertalanffy, L. von: Theoretısche Biologie. Vol. 2. Bern: A. Francke Verlag, 1951

Bertalanffy, L. von: Wachstum. In: Kükenthal, Handbuch der Zoologıe, Vol. 8, 10. Lieferung, Teil 4 (6), S. 1–68 Berlin: Walter de Gruyter 1957

Björkman, N.: An atlas of placental fine structure. London: Baillière, Tındall & Cassel 1970

Brown, A. L.: Microvilli of the human jejunal epithelial cell. J. Cell Biol. 12, 623–627 (1962)

Carlyle, A.: An ıntegration of the total oxygen consumption of the sheep fetus from that of the tissues. J. Physiol. 107, 355–364 (1948)

Cattoor, J. P.: Evolution de la surface villosıtaire d'échange. In: Bret, J.: Les avortements spontanés du premier trimestre et le placenta. pp. 33–62. Paris: Masson 1967

Comline, R. S., Silver, M.: Placental transfer of blood gases. Brit. med. Bull. 31, 25–31 (1975)

Dick, E. G., Dıck, D. A. T., Bradbury, S.: The effect of surface microvilli on the water permeability of single toad oocytes. J. Cell Sci. 6, 451–476 (1970)

Fawcett, D. W.: The Cell, an atlas of fine structure. Philadelphia-London: Saunders 1966

Franke, H.: Feinstruktur der Placenta. Jena: Fischer 1969

Fujikura, T., Ezaki, K., Nishimura, H.: Chorionic villi and syncytial sprouts in spontaneous and ınduced abortions. Amer. J. Obstet. Gynec. 110, 547–555 (1971)

Gehr, P., Bachofen, M., Weibel, E. R.: Elektronenmikroskopie und Morphometrie der menschlichen Lunge. Vortrag an der 37. Tagung der freien Vereinigung der Anatomen an Schweizerischen Hochschulen (1974). Adresse der Autoren: Anatomisches Institut der Universität Bern

Geissler, U., Holtorff, J., Hempel, K.: Morphometrische Studien an der Placenta. Teil I: Die Grösse der Zottenoberfläche am Ende der normal verlaufenden Schwangerschaft. Zbl. Gynäk. 94, 4–11 (1972)

Grosser, O.: Vergleichende Anatomie und Entwicklungsgeschichte der Eihäute und der Placenta mit besonderer Berücksichtigung des Menschen. Wien-Leipzig: Braumüller 1909

Gurlt, E. F.: Handbuch der vergleichenden Anatomie der Haussäugetiere, 5. Aufl. Berlin: Hırschwald 1873

Hörmann, G.: Lebenskurven normaler und entwicklungsunfähiger Chorionzotten. Arch. Gynäk. 181, 29–43 (1951)

Huggett, A. St. G., Widdas, W. F.: The relationship between mammalian foetal weight and conception age. J. Physiol. **114**, 306–317 (1951)

Jakobovits, A., Iffy, L., Wingate, M. B., Slate, W. G., Chatterton, R. T., Kerner, P.: The rate of early fetal growth in the human subject. Acta anat. (Basel) **83**, 50–59 (1972)

Janisch, E.: Das Exponentialgesetz als Grundlage einer vergleichenden Biologie. Abhandlungen zur Theorie der organischen Entwicklung, Heft 2 (= Neue Folge von Roux' Vorträgen und Aufsätzen über Entwicklungsmechanik der Organismen). Berlin: Springer 1927

Janssen, A.: Über die Placenta und Paraplacenta bei der Katze. Diss. Hannover 1933

Jewett, D. A., Dukelow, W. R.: Cyclicity and gestation length of Macaca fascicularis. Primates **13**, 327–330 (1972)

Kirchshofer, R.: Gorillazucht in Zoologischen Garten und Forschungsstationen. Der Zoologische Garten **38**, Heft 3/4 (1970)

Kleiber, M.: Body size and metabolic rate. Physiol. Rev. **27**, 511–541 (1947)

Kleiber, M.: Der Energiehaushalt von Mensch und Säugetier. Hamburg-Berlin: Parey 1967

Knopp, J.: Das Wachstum der Chorionzotten vom II. bis X. Monat. Z. Anat. Entwickl. Gesch. **122**, 42–59 (1960)

Koller, S.: Typisierung korrelativer Zusammenhange. Metrika **6**, 65–75 (1963)

Kretschmann, H.-J., Wingert, F.: Biometrische Analyse der Volumina des Striatum einer ontogenetischen Reihe von Albinomausen. Z. Anat. Entwickl.-Gesch. **128**, 85–108 (1969)

Laga, E. M., Driscoll, S. G., Munro, H. N.: Quantitative studies of human placenta. I. Morphometry Biol. Neonate **23**, 231–259 (1973)

Lehmann, G.: Das Gesetz der Stoffwechselreduktion in der höheren Tierwelt. Z. Naturforsch. **6b**, 216–223 (1951)

Ludwig, K. S.: Zur vergleichenden Histologie des Allantochorion. Rev. suisse Zool. **75**, 819–831 (1968)

Marrable, A. W.: The embryonic pig. London-New York: Pitman 1971

Mehta, V. S., Prakash, A. H. A., Singh, M.: Gestation period in camels. Indian Vet. J. **39**, 387–389 (1962)

Meschia, G.: Normal exchange of respiratory gases across the sheep placenta. In: Longo, L. D. and Bartels, H.: Respiratory gas exchange and blood flow in the placenta. U. S. Department of Health, Education and Welfare, DHEW Publication No (NIH) 73–361, p. 229–238, Bethesda/Maryland 1972

Michel, G.: Embryologie. In: Schwarze, E. und Schröder, L. Kompendium der Veterinär-Anatomie, Bd. 6. Jena: Fischer 1968

Moll, W.: Gas exchange in concurrent, countercurrent and crosscurrent flow systems. In: Longo, L. D. and Bartels, H.: Respiratory gas exchange and blood flow in the placenta (DHEW Publication No (NIH) 73–361) p. 281–296. Bethesda/Maryland 1972

Müller, G. A., Wernicke, E. H., Fischer, W. M.: Morphometrische Untersuchungen am Labyrinth der Meerschweinchenplazenta. In: Elias, H.: Stereology, Proc. 2nd Int. Congr. Stereology, Chicago 1967. Berlin-Heidelberg-New York: Springer 1967

Naaktgeboren, C.: Das embryonale Wachstum des Rindes mit besonderer Berücksichtigung der fur die Geburt wichtigen Körperteile. Z. Morph. Oekol. Tiere **48**, 447–460 (1960)

Naaktgeboren, C. und Slijper, E. J.: Biologie der Geburt. Hamburg-Berlin: Parey 1970

Needham, J.: Chemical embryology. New York-London: Hafner 1963

Ostwald, W.: Ueber die zeitlichen Eigenschaften der Entwicklungsvorgange. In: Roux, W.: Vortrage und Aufsatze über Entwickelungsmechanik der Organismen, Heft 5. Leipzig: Engelmann 1908

Peil, J.: Mathematische Beschreibung von Wachstumsvorgängen. Morph. Jb. **120**, 832–853 (1974)

Portmann, A.: Die Ontogenese der Saugetiere als Evolutionsproblem. Biomorphosis **1**, 49–66 (1938)

Postma, C.: De ouderdomsbepaling bij Runderfoetus. T. Diergeneesk. **72**, 463–531 (1947)

Sachs, L.: Statistische Auswertungsmethoden, 3. Aufl. Berlin-Heidelberg-New York: Springer 1972

Scharf, J. H.: Differentialgleichungen in der funktionellen Morphologie. Gegenbaurs morph. Jb. **117**, 3–38 (1971)

Slijper, E. J.: Riesen und Zwerge im Tierreich. Hamburg-Berlin: Parey 1967

Sokal, R. R., Rohlf, F. J.: Biometry. San Francisco: Freeman 1969

Starck, D.: Embryologie. Stuttgart: Thieme 1975

Stegeman, J. H. J.: Placental development in the sheep and its relation to fetal development. Bijdragen tot de dierkunde **44**, 1–72 (1974)

Stoss, A. O.: Tierarztliche Geburtskunde und Gynakologie, 2. Aufl. Stuttgart: F. Enke 1944

Strahl, H.: Die Embryonalhüllen der Säuger und die Placenta. In: Hertwig's Handbuch der vergleichenden und experimentellen Entwicklungslehre der Wirbeltiere, Bd. 1_2. pp. 235–368. Jena: Fischer 1906

Streeter, G. L.: Weight, sitting heigh, head size, foot lenght and menstrual age of the human embryo. Publ. No. 274. Contr. Embryol. Carneg. Instn. **11**, 143–170 (1920)

Taverne, M. A. M., and Bakker-Slotboom, M. F.: Observations on the delivered placenta and fetal membranes of the Aardvark, Orycteropus afer (Pallas, 1766). Bijdragen tot de dierkunde **40** (2), 154–162 (1970)

Teissier, G.: La relation d'allométrie, sa signification statistique et biologique. Biometrics **4**, 14–53 (1948)

Underwood, E. E.: Quantitative Stereology. Reading, Massachusetts: Addison-Wesley 1970

Villee, C. A.: The placenta and fetal membranes. Baltimore: Williams and Wilkins 1960

Weibel, E. R.: Stereological principles for morphometry in electron microscopic cytology. Int. Rev. Cytol. **26**, 235–302 (1969)

Widdowson, E. M.: Growth and composition of the fetus and newborn. In: Assali, N. S.: Biology of gestation. New York-London: Academic press 1968

Zietzschmann, O., Krölling, O.: Lehrbuch der Entwicklungsgeschichte der Haussäugetiere, 2. Aufl. Hamburg-Berlin: Parey 1955

Subject Index

The numbers set in *italics* refer to pages where the term appears in a figure or a table

Other Reviews of Interest in this Series

Part 4: **Ribi, W. A.**: The Neurons of the First Optic Ganglion of the Bee (Apis mellifera). 21 figures. 43 pages. 1975. ISBN 3-540-07096-6

Part 5: **Halata, Z.**: The Mechanoreceptors of the Mammalian Skin. Ultrastructure and Morphological Classification. 11 figures. 77 pages. 1975. ISBN 3-540-07097-4

Part 6: **Beckers, H. W.; Eisenacher, W.**: Zur Morphologie der Papilla fungiformis einiger Primaten und des Menschen. Zur Morphologie der Papilla fungiformis einiger Nagetiere. Rasterelektronenmikroskopische, licht- und elektronenmikroskopische Untersuchungen. 27 figures. 117 pages. 1975. ISBN 3-540-07098-2

Volume 51

Part 1: **Putte, S. C. J. van der**: The Development of the Lymphatic System in Man. 33 figures. 60 pages. 1975. ISBN 3-540-07204-7

Part 2: **Raedler, A., Sievers, J.**: Influences of Experimental Brain Edema on the Development of the Visual System. 27 figures. 60 pages. 1975. ISBN 3-540-07205-5

Part 3: **Pexieder, T.**: Cell Death in the Morphogenesis and Teratogenesis of the Heart. 52 figures. 100 pages. 1975. ISBN 3-540-07270-5

Part 4: **Svendgaard, N. A.; Björklund, A.; Stenevi, U.**: Regnerative Properties of Central Monoamine Neurons. 24 figures. 77 pages. 1975. ISBN 3-540-07299-3

Part 5: **Gossrau, R.**: Die Lysosomen des Darmepithels. 74 figures. 95 pages. 1975. ISBN 3-540-07271-3

Part 6: **Thorn, L.**: Die Entwicklung des Cortischen Organs beim Meerschweinchen. 23 figures. 97 pages. 1975. ISBN 3-540-07301-9

Volume 52

Part 1: **Ibrahim, M. Z. M.**: Glycogen and its Related Enzymes of Metabolism in the Central Nervous System. 13 figures. 89 pages. 1975. ISBN 3-540-07454-6

Part 2: **Cau, P.; Michel-Béchet, M.; Fayet, G.**: Morphogenesis of Thyroid Follicles in Vitro. 16 figures. 66 pages. 1976. ISBN 3-540-07654-9

Part 3: **Tiedemann, K.**: The Mesonephros of Cat and Sheep. Comparative Morphological and Histochemical Studies. 47 figures. 119 pages. 1976. ISBN 3-540-07779-0

Part 4: **Haug, F.-M. Š.**: Sulphide Silver Pattern and Cytoarchitectonics of Parahippocampal Areas in the Rat. Special Reference to the Subdivision of Area Entorhinalis (Area 28) and its Demarcation from the Pyriform Cortex. 49 figures. 73 pages. 1976. ISBN 3-540-07850-9

Part 5: **Phillips, I. R.**: The Embryology of the Common Marmoset (Callithrix jacchus). 22 figures. 47 pages. 1976. ISBN 3-540-07955-6

Part 6: **Nobiling, G.**: Die Biomechanik des Kieferapparates beim Stierkopfhai. 25 figures. 52 pages. 1977. ISBN 3-540-08038-4

Springer-Verlag Berlin · Heidelberg · New York